AF358726

Advances in Methane Production from Coal, Shale and Other Tight Rocks

Advances in Methane Production from Coal, Shale and Other Tight Rocks

Editors

Yong Li
Fan Cui
Chao Xu

MDPI • Basel • Beijing • Wuhan • Barcelona • Belgrade • Manchester • Tokyo • Cluj • Tianjin

Editors
Yong Li
China University of Mining
and Technology (Beijing)
China

Fan Cui
China University of Mining
and Technology (Beijing)
China

Chao Xu
China University of Mining
and Technology (Beijing)
China

Editorial Office
MDPI
St. Alban-Anlage 66
4052 Basel, Switzerland

This is a reprint of articles from the Special Issue published online in the open access journal *Energies* (ISSN 1996-1073) (available at: https://www.mdpi.com/journal/energies/special_issues/Methane_Production_Coal_Shale).

For citation purposes, cite each article independently as indicated on the article page online and as indicated below:

LastName, A.A.; LastName, B.B.; LastName, C.C. Article Title. *Journal Name* **Year**, *Volume Number*, Page Range.

ISBN 978-3-0365-6003-8 (Hbk)
ISBN 978-3-0365-6004-5 (PDF)

Contents

About the Editors

Yong Li

Yong Li, from the College of Geosciences and Surveying, China University of Mining and Technology (Beijing), is an associate professor in petroleum and natural gas geology. His interests are in the areas of sedimentology, reservoir characterization, and issues associated with unconventional hydrocarbons, including coalbed methane, tight gas, shale gas, and gas hydrate.

Fan Cui

Fan Cui, from the School of Geosciences and Surveying Engineering, China University of Mining and Technology (Beijing), is a professor in geophysical prospecting and information technology. His interests are in the areas of applied geophysics, ground penetrating radar method and technology, 3D seismic inversion of coalfield, and basic research on transparent mine technology. Dr. Cui has published more than 50 technical papers in professional journals and conference proceedings.

Chao Xu

Chao Xu, from the School of Emergency Management and safety Engineering, China University of Mining and Technology (Beijing), is an associate professor in safety science and engineering. His interests are in the areas of coal mine safety, coal and rock dynamic disaster prevention, and issues associated with methane extractions, including coalbed methane, coal mine methane, and abandoned mine methane.

Preface to "Advances in Methane Production from Coal, Shale and Other Tight Rocks"

Unconventional natural gas, including shale gas, tight gas, and coalbed methane, looks set to have an important role in the present and future clean energy structure. Technological advancements in horizontal drilling and hydraulic fracturing have spurred a rapid increase in unconventional energy production over the past decade. The application of new technologies has enabled natural gas and shale oil to be economically produced from shale and other unconventional formations. Further, various methods have been adopted to improve gas recovery, including but not limited to high-precision characterization of coal and shale reservoirs at multiscales, fast drilling and completion of wells with long laterals as well as large-scale volume fracturing. Numerous new techniques have emerged, such as multiple well-type development, fluid injection, nano-flooding, and enhancing biogenic methane generation methods.

This collection reports on the state of the art in fundamental discipline application of these techniques to methane production and associated challenges in geoengineering activities. Zheng et al. (2022) report on an NMR-based method for multiphase methane characterization in coals, which is helpful for understanding methane occurrence within deep coals. Wang et al. (2022) studied the genesis of bedding fractures in Ordovician to Silurian marine shale in the Sichuan basin, China, which is significant for shale gas production. Kang et al. (2022) proposed research focusing on the differentiation and prediction of shale gas production from horizontal wells, which can be applied in the Weiyuan Shale Gas Field, Sichuan Basin, China. Liang et al. (2022) studied the pore structure of marine shale from the terms of molecular interaction via the adsorption method. Zhang et al (2022) focus on the coal measures sandstone reservoir in Xishanyao Formation, at the Southern Margin of the Junggar Basin, and its sandstone diagenetic characteristics are fully revealed. As with the exploration on marine areas, Yao et al. (2022) report on the source-to-sink system in the Ledong submarine channel and the Dongfang submarine fan in the Yinggehai Basin, South China Sea.

There are 4 papers focusing on the technologies associated with hydrocarbon production from the tight rocks. Wang et al. (2022) report the analysis of pre-stack inversion in carbonate karst reservoir, which shows good application potential. Chen et al. (2022) conducted an inversion study on parameters of cascade coexisting gas-bearing reservoirs in coal measures in Huainan, south China, which is useful in the coal measure gas accumulation prediction and production. To ensure the safety of carbon capture and storage (CCS), Zhang et al (2022) report their analysis of available conditions for InSAR surface deformation monitoring in CCS projects. Additionally, to ensure the safety of coal mine production, Zhang et al. (2022) report the properties and applications of gel materials for coal gangue control.

All 10 papers featured focus on successful unconventional hydrocarbon production. We hope that the collection can be beneficial for readers in the field of unconventional oil and gas geology, and also to energy-related scholars.

Acknowledgments: We thank the representative colleges of the China University of Mining and Technology (Beijing) for supporting the compilation of this Special Issue. Editor Qisheng Su helped to coordinate this Special Issue.

Yong Li, Fan Cui, and Chao Xu

Editors

Article

An NMR-Based Method for Multiphase Methane Characterization in Coals

Sijian Zheng [1,2], Shuxun Sang [1,2,3,*], Shiqi Liu [1,2,*], Xin Jin [4,5], Meng Wang [1,2], Shijian Lu [1,2], Guangjun Feng [3], Yi Yang [4] and Jun Hou [6]

[1] Carbon Neutrality Institute, China University of Mining and Technology, Xuzhou 221008, China; sijian.zheng@cumt.edu.cn (S.Z.); wangm@cumt.edu.cn (M.W.); lushijian@cumt.edu.cn (S.L.)

[2] The Key Laboratory of Coal-Based CO_2 Capture and Geological Storage, China University of Mining and Technology, Xuzhou 221116, China

[3] School of Resources and Geosciences, China University of Mining and Technology, Xuzhou 221116, China; gjfeng@cumt.edu.cn

[4] Xi'an Research Institute, China Coal Technology and Engineering Group, Xi'an 710077, China; jinxin@cctegxian.com (X.J.); yangyi@cctegxian.com (Y.Y.)

[5] Key Laboratory of Metallogenic Prediction of Nonferrous Metals and Geological Environment Monitoring, Ministry of Education, School of Geosciences and Info-Physics, Central South University, Changsha 410083, China

[6] College of Civil Engineering, Anhui Jianzhu University, Hefei 230601, China; houjun2132234641@163.com

* Correspondence: shxsang@cumt.edu.cn (S.S.); liushiqi@cumt.edu.cn (S.L.)

Citation: Zheng, S.; Sang, S.; Liu, S.; Jin, X.; Wang, M.; Lu, S.; Feng, G.; Yang, Y.; Hou, J. An NMR-Based Method for Multiphase Methane Characterization in Coals. *Energies* 2022, *15*, 1532. https://doi.org/10.3390/en15041532

Academic Editor: Nikolaos Koukouzas

Received: 9 January 2022
Accepted: 16 February 2022
Published: 18 February 2022

Publisher's Note: MDPI stays neutral with regard to jurisdictional claims in published maps and institutional affiliations.

Abstract: Discriminating multiphase methane (adsorbed and free phases) in coals is crucial for evaluating the optimal gas recovery strategies of coalbed methane (CBM) reservoirs. However, the existing volumetric-based adsorption isotherm method only provides the final methane adsorption result, limiting real-time dynamic characterization of multiphase methane in the methane adsorption process. In this study, via self-designed nuclear magnetic resonance (NMR) isotherm adsorption experiments, we present a new method to evaluate the dynamic multiphase methane changes in coals. The results indicate that the T_2 distributions of methane in coals involve three different peaks, labeled as P1 ($T_2 < 8$ ms), P2 ($T_2 = 20$–300 ms), and P3 ($T_2 > 300$ ms) peaks, corresponding to the adsorbed phase methane, free phase methane between particles, and free phase methane in the sample cell, respectively. The methane adsorption Langmuir volumes calculated from the conventional volumetric-based method qualitatively agree with those obtained from the NMR method, within an allowable limit of approximately ~6.0%. Real-time dynamic characterizations of adsorbed methane show two different adsorption rates: an initial rapid adsorption of methane followed by a long stable state. It can be concluded that the NMR technique can be applied not only for methane adsorption capacity determination, but also for dynamic monitoring of multiphase methane in different experimental situations, such as methane adsorption/desorption and CO_2-enhanced CBM.

Keywords: low-field NMR; coal; free methane; paramagnetic mineral

1. Introduction

The increasing attention paid to clean energy has promoted the efficient development and exploration of the coalbed methane (CBM) industry [1–4]. At the same time, the scientific characterization of coals' petrophysical properties, such as pore structure, permeability, and methane adsorption capacity, have recently aroused great research interest [5–8]. For example, the methane adsorption capacity is a **crucial parameter** in evaluating the methane content of CBM reservoirs [7,9–11]. Additionally, investigating the methane adsorption process in coal pore systems is essential to establish and assess CBM wells' production [12,13]. Thus, an accurate and real-time dynamic characterization of the methane adsorption process in coals is necessary for CBM production.

Commonly, the methane adsorption capacity is determined using the volumetric-based adsorption isotherm [14,15]. However, this method is quite complicated and susceptible to impurities or the volume effect, may result in larger variations in experimental results [16]. Furthermore, the Langmuir isotherm curve consists only of cross-plots of adsorbed methane content and experimental pressure, and has the significant shortcoming that it is incapable of depicting any details in the methane adsorption process. The dynamic characterization of the methane adsorption process in coals is usually evaluated based on the assumption of a single pore or regular structure pore network model, which is derived from well-established sandstone or tight sandstone [17,18]. However, due to complicated and heterogeneous coal pore structures, the model obtained from sandstone or tight sandstone needs to be modified, which is complex and time consuming. To the best of the authors' knowledge, few methods can be individually and directly applied for the dynamic characterization of methane adsorption in coals.

Nuclear magnetic resonance (NMR) is a non-destructive technique that has been widely applied for investigating hydrogen-bearing reservoir fluids' petrophysical properties [19–25]. Based on the fully water-saturated and centrifuged NMR transverse relaxation time (T_2) distribution of coals, Zheng et al. [22] calculated the T_2 cutoff values and then used them to classify pore types and evaluate the full-scale pore size distribution. Guo et al. [26] first applied the NMR T_2 measurement to evaluate the methane T_2 distributions of low-rank coals. They found three different methane relaxation mechanisms existing in coals: adsorbed methane, porous medium confined methane, and free methane. Vasilenko [27] presented an NMR study of the ratio between the free and adsorbed phase methane in fossil coals, and found that the adsorbed methane was the predominant phase state, only upon the opening of a high-pressure chamber after the emission of methane from filtration channels. Liu et al. [28] discussed the CO_2 enhanced gas recovery efficiency in shales based on NMR measurement. They found that the higher concentration ratios of CO_2/CH_4 are more efficient for gas recovery. By introducing the NMR relaxation method, Yao et al. [29] investigated the multiphase methane relaxation characterization in shales. In their study, NMR peaks with $T_2 < 1$ ms and 1–50 ms corresponded to the adsorbed phase methane and free phase methane in shale, respectively. Additionally, Yao et al. [30] characterized the methane adsorption capacity on two low volatile bituminous coals using low-field NMR measurements and revealed that the P1 peak amplitude increases rapidly at the beginning stages, and then trends to attain an ultimate value, in a manner similar to the Langmuir equation. However, the research objects in Yao et al. [30] were only two low-rank coals, which restricted the NMR relaxation measurement application for medium- and high-rank coals, especially for some coals containing paramagnetic minerals. Previous achievements have verified the qualitative ability of the NMR relaxation method in methane adsorption characterization in unconventional reservoirs (e.g., shale and coal). Moreover, few studies relate to the application of the NMR data for estimating the real-time dynamic methane adsorption process in coals.

In this paper, a series of comparative volumetric- and NMR-based methane adsorption measurements were performed for eight coals whose ranks were strikingly different. Comparing experimental results from the volumetric method, the accuracy of the NMR-based method for methane adsorption characterization in coals having different ranks was validated. Based on the NMR T_2 distribution in different time intervals, the methane adsorption dynamic process characteristics in coals were investigated.

2. Experimental Methods

2.1. Coal Sampling

In this study, a total of eight block coals ($20 \times 20 \times 20$ cm^3) having different ranks were collected from the Junggar basin, Ordos basin, and Qinshui basin, China. Table 1 shows the summary of the detailed basic petrophysical coal information. The maximum vitrinite reflectance ($R_{o,m}$) of the selected samples ranges from 0.52% to 3.03%, covering a wide array of coal ranks, including low, medium, and high ranks.

Table 1. Detailed information of the selected coals.

Sample No.	$R_{o,m}$ (%)	Maceral Composition (%)				Proximate Analysis (%)		
		V	I	E	M	M_{ad}	A_d	FC_d
L1	0.52	64.3	30.7	4.7	0.3	7.67	21.34	44.95
L2	0.60	64.8	16.1	16.4	2.7	5.82	11.15	45.83
L3	0.70	50.2	37.2	4.7	7.9	2.55	15.49	56.42
M1	1.52	50.9	20.3	19.1	9.7	5.43	33.54	33.77
M^2	1.68	58.6	22.1	14.3	5.0	6.30	29.19	36.51
H1	2.36	80.3	10.7	1.0	8.0	0.88	17.8	87.21
H2	2.54	83.4	15.9	0	0.7	0.74	12.06	75.05
H3	3.03	86.2	10.1	0	3.7	1.46	1.62	95.94

Notes: $R_{o,m}$—maximum vitrinite reflectance, V—vitrinite; I—inertinite; E—exinite; M—minerals. M_{ad}—moisture (air-dried basis); A_d—ash (dry basis); FC_d—carbon (air-dried basis).

Figure 1 shows the pore morphology characteristics of the selected coals by the scanning electron microscope (SEM) measurement. Results show that the pore types developed in samples L3, M1, H1, and H2 are mainly gas pores. Intergranular pores and residual plant tissue pores are found in samples M2 and H3, respectively, whereas fractures are well developed in samples L1 and L2.

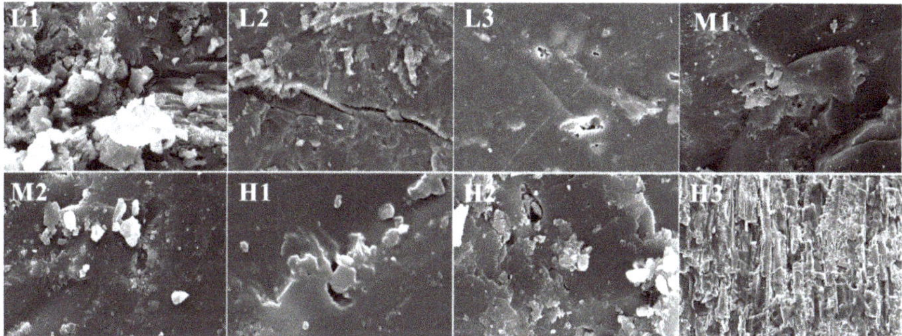

Figure 1. Pore morphology characteristics of the selected coals by SEM measurement.

2.2. NMR Adsorption Measurements

Compared with NMR T_1 measurements, T_2 measurements are preferred for the characterization of petrophysical properties in rocks due to their obvious advantages, such as short testing time, good applicability, and convenient operation. Generally, the NMR T_2 characteristics are tested by applying the CPMG sequence. Based on the NMR principle, T_2 can be characterized as follows [31–35]:

$$\frac{1}{T_2} = \frac{1}{T_{2B}} + \frac{1}{T_{2S}} + \frac{1}{T_{2D}} = \frac{3T_k}{298\eta} + \rho_2\left(\frac{S}{V}\right) + \frac{D(\gamma GT_E)^2}{12} \tag{1}$$

where T_{2B} represents bulk relaxation; T_{2S} represents surface relaxation; T_{2D} represents diffuse relaxations; T_k means experimental temperature; η means methane viscosity; ρ_2 means surface relaxivity of the sample; S means specific surface area; V means pore volume; D means methane diffusion coefficient; γ means proton gyromagnetic ratio; G means magnetic field intensity; T_E means echo spacing time.

$$\frac{1}{T_2} = \frac{1}{T_{2B}} + \rho_2\left(\frac{S}{V}\right) \tag{2}$$

As shown in Figure 2, the experimental set-up of NMR methane adsorption measurement mainly includes five parts: (1) the gas supply system, providing a different gas source for the experimental system; (2) the gas exhaust system, recycling exhaust gas after experiments; (3) a non-magnetic PEEK sample cell, for placement of coal powders; (4) a non-magnetic PEEK reference cell, for transporting and sustaining methane pressure for the sample cell; and (5) an NMR measurement device, measuring the NMR T_2 distributions in methane adsorption process. Here, NMR experimental parameters were set to: waiting time, 3000 ms; echo spacing, 0.3 ms; echo numbers, 10,000.

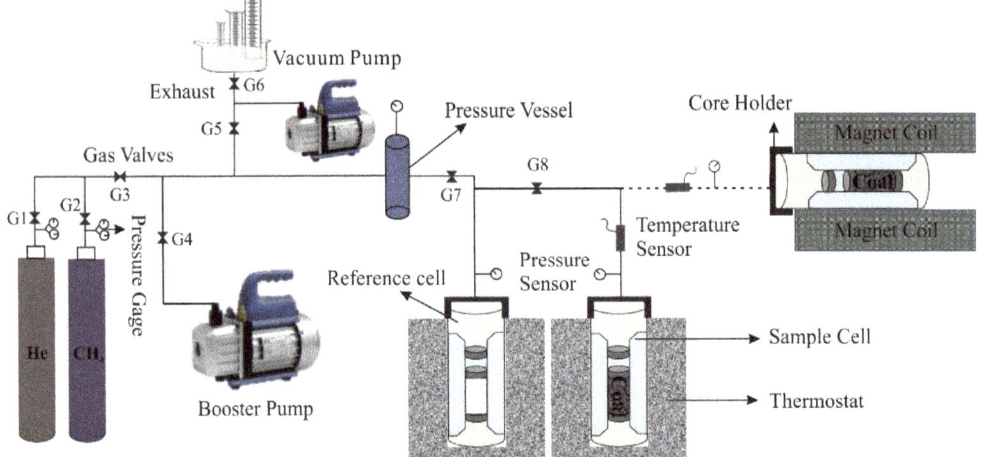

Figure 2. Schematic diagram of the NMR methane adsorption experimental set-up.

Two independent experiments were performed for methane adsorption characterization by the NMR measurement; one was a free methane T_2 relaxation property experiment (no powder coals in the sample cell) and the other was a methane adsorption T_2 relaxation experiment.

The free methane T_2 relaxation property was measured under different methane pressures, at a stable temperature (304.15 K). The amount of free methane can be determined according to the ideal-gas equation. At the same time, the NMR relaxation distributions of free methane under different pressures were measured using an NMR device.

Prior to methane adsorption T_2 relaxation experiments, the sample cell and reference cell must be cleaned to prevent the contamination of impurities. Then, 60–80 mesh powder coals were dried at 368 K in a vacuum oven at least for one day. The workflow of methane adsorption T_2 relaxation experiments was as follows:

P1: The powder coals were transformed into the sample cell and placed in vacuum for three hours;

P2: ~8 MPa free methane was injected into the reference cell;

P3: The G8 valve was switched on to introduce the reference cell methane into the sample chamber until it reached the set pressure;

P4: The NMR T_2 distributions were continuously measured for methane adsorption until completely adsorbed, with an interval time of 45 min;

P5: The sample cell methane pressure was increased to six different pressure levels, which were repeated for the same sample.

3. Results and Discussions

3.1. Free Methane T_2 Distributions

As shown in Figure 3, the T_2 characteristics for free methane under different pressures exhibit one significant peak. This peak appears at approximately ~50–1500 ms. As shown in Figure 4, the NMR spectra amplitude shows an evident linear relationship with the free methane mass:

$$y = 0.000015x \left(R^2 = 0.9990 \right) \tag{3}$$

where y means the free methane mass; x means the free methane NMR signal amplitude.

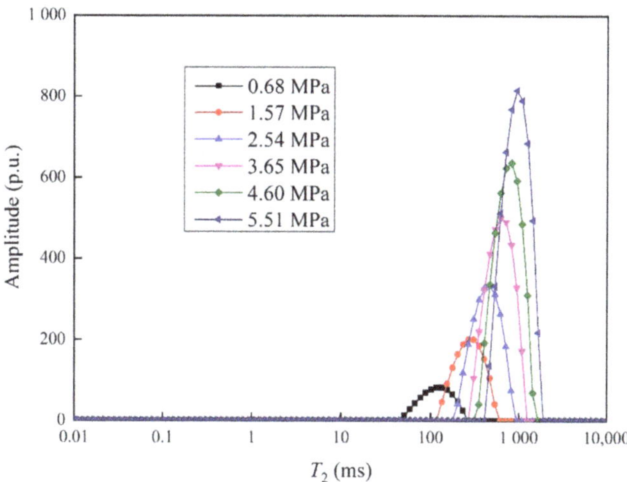

Figure 3. NMR T_2 distributions of free methane.

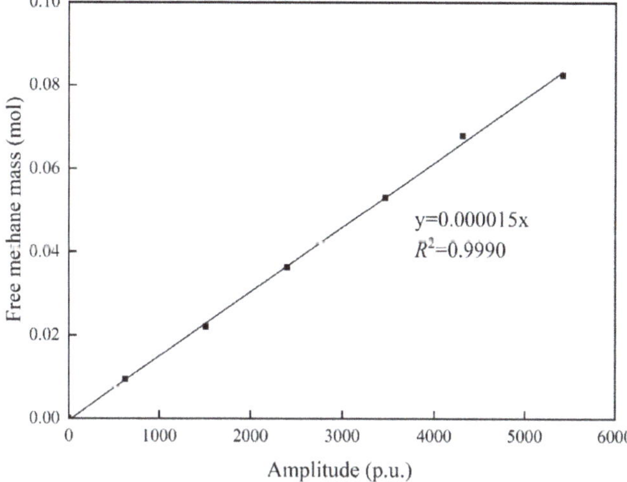

Figure 4. Correlations between T_2 amplitude vs. methane mass.

3.2. Methane Adsorption T_2 Relaxation Characteristics

Methane adsorption T_2 spectra for the selected coals are presented in Figure 5. The minimal T_2 amplitude of the dry samples (solid purple line) indicates the nuclear responses from the coal itself can be ignored. Results show that the methane adsorption T_2 spectra exhibit three peaks: P1, P2, and P3. The NMR P1 peak appears in approximately ~0.1–8 ms, whereas the P2 and P3 peaks emerge in approximately ~20–300 ms and 300–2000 ms, respectively. Based on the NMR principle in Equation (2) and the research results in Yao et al. [30], the multiphase methane in coals characterized by NMR was divided into three parts: (1) the adsorbed phase occurring in the coal pore surface, represented by the NMR P1 peak; (2) the free phase emerging between coal particles, represented by the NMR P2 peak; and (3) the free phase occurring in the sample cell free space, represented by the NMR P3 peak. Here, we defined the multiphase methane in coals as the adsorbed phase and the free phase.

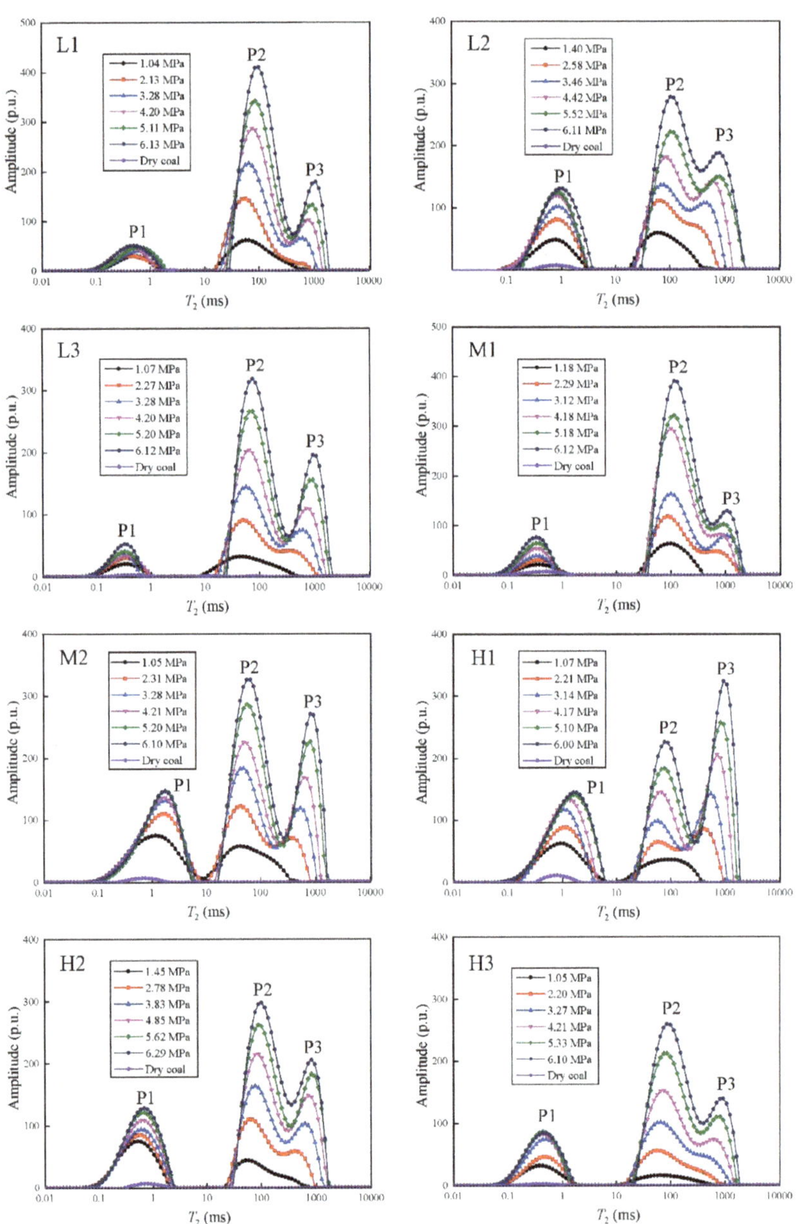

Figure 5. NMR T_2 spectra of methane adsorption in coals.

Figure 5 shows that all peak (P1, P2, and P3) amplitudes have an increasing trend with the increase in methane pressure. In order to advance quantitative characterization of the

NMR P1 peak variation, the relationship between NMR P1 peak amplitude and methane pressure is plotted in Figure 6a. The results show that the NMR P1 peak amplitude rapidly increases at the beginning stages, and then trends to attain an ultimate value, in a manner similar to the Langmuir equation.

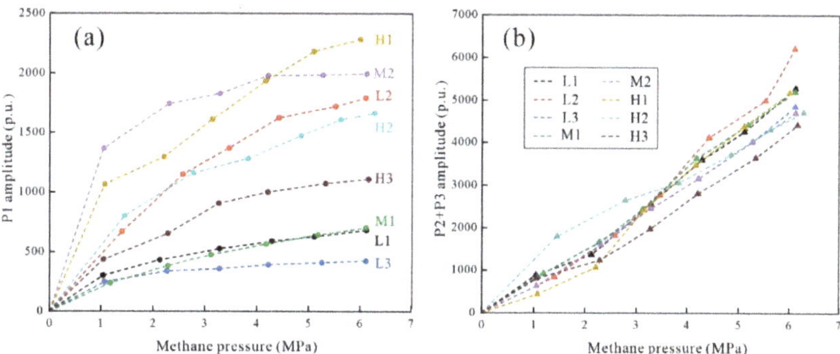

Figure 6. NMR P1 (**a**) and P2 + P3 (**b**) amplitude changes during the process of methane adsorption.

Figure 6b shows that the free phase methane amplitudes have a linear relationship with the methane pressure. The results in Figure 6 confirm that the NMR P1 peak represents the adsorbed phase methane, and NMR P2 + P3 represent the free phase methane. Moreover, the adsorbed methane content cannot be determined by NMR P1 amplitude directly because the relaxation mechanisms are obviously different between free phase methane and adsorbed phase methane. Free phase methane was controlled by bulk relaxation, whereas adsorbed phase methane surface relaxation was controlled by surface relaxation [36].

3.3. Methane Adsorption Capacity Determination

Methane adsorption capacity is **crucial for** understanding and predicting the CBM content [37–40]. This section provides a detailed comparison of the methane adsorption capacity based on the volumetric method and the self-designed NMR method. Then, the application of the NMR-based method to different coals is further estimated and discussed.

3.3.1. Volumetric-Based Method

Volumetric-based adsorption measurements are commonly used for calculating the methane adsorption capacity in coals. Based on the equation of the state of an ideal gas, the reduced amount of free phase methane can be calculated based on the reduction in sample cell pressure:

$$PV' = nZRT \tag{4}$$

where P means experimental pressure; V' means free methane volume; n means free methane amount; Z means compressibility factor; R is 8.3144 J/(mol·K); T means experimental temperature.

Based on the volumetric-based adsorption experimental data, including adsorption pressure (P) and adsorption volume (V), the Langmuir volume (V_L) and pressure (P_L) can be calculated using the Langmuir equation:

$$V = \frac{V_L P}{P + P_L} \tag{5}$$

where V means experimental adsorption volume, equaling to the reduction in free methane volume in the cells; P means experimental pressure, which can be obtained directly from a pressure sensor in the sample cell. V_L and P_L mean the Langmuir volume and Langmuir pressure, which can be determined based on Langmuir curve fitting between the parameters of experimental adsorption volume (V) and experimental pressure (P).

As shown by the blue dotted line in Figure 7, the volumetric-based adsorption experimental data exhibit an excellent fit with the Langmuir curve (Figure 7). As listed in Table 2, the volumetric-based Langmuir volumes ($V_{L\text{-vol}}$) are in the range of 5.79–21.14 cm^3/g, whereas the volumetric-based Langmuir pressure ($P_{L\text{-vol}}$) ranges from 0.65 to 2.75 MPa.

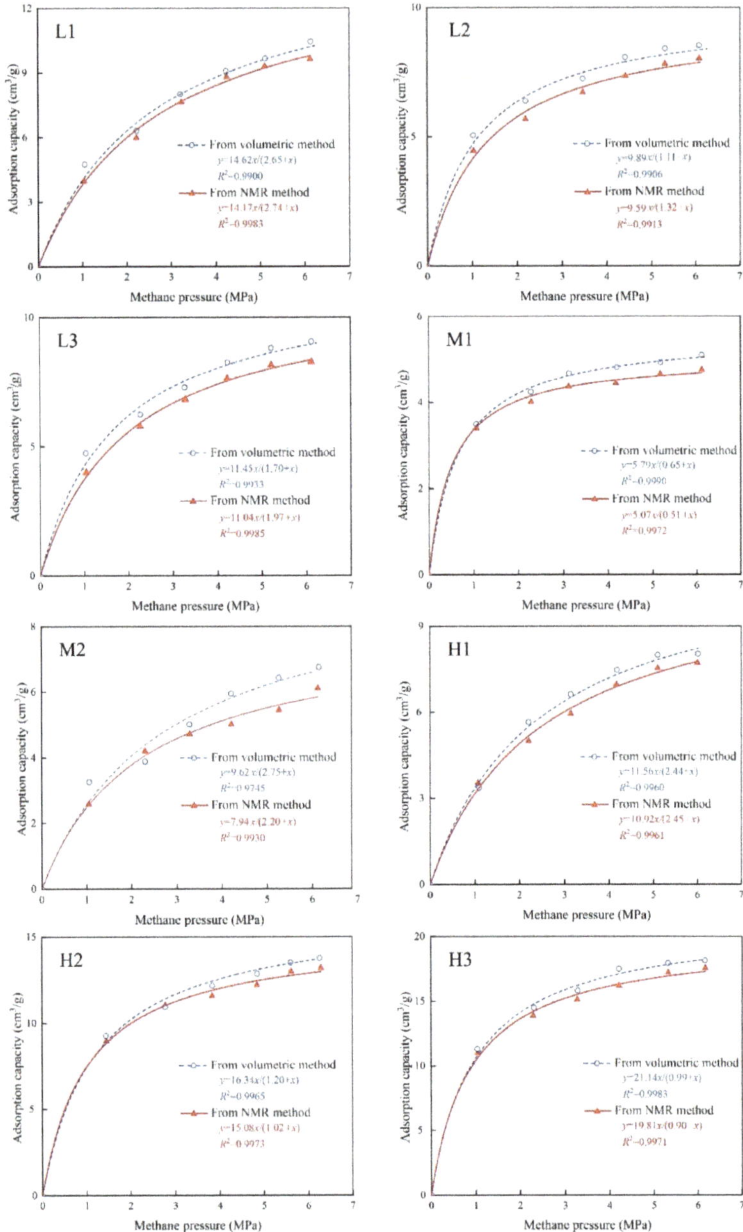

Figure 7. Adsorption isotherms obtained from two different methods.

Table 2. Comparison of the Langmuir parameters determined by volumetric-based and NMR-based methods.

Sample No.	Volumetric Isotherm Adsorption Method			NMR Isotherm Adsorption Method			Powder Mass (g)
	$V_{L\text{-vol}}$ (cm^3/g)	$P_{L\text{-vol}}$ (MPa)	Adjusted R-Square	$V_{L\text{-NMR}}$ (cm^3/g)	$P_{L\text{-NMR}}$ (MPa)	Adjusted R-Square	
L1	14.62	2.65	0.9900	14.17	2.74	0.9983	14.6
L2	9.89	1.11	0.9906	9.59	1.32	0.9913	16.5
L3	11.45	1.70	0.9933	11.04	1.97	0.9985	17.3
M1	5.79	0.65	0.9990	5.07	0.51	0.9972	12.9
M2	9.62	2.75	0.9745	7.97	2.20	0.9930	18.9
H1	11.56	2.44	0.9960	10.92	2.45	0.9961	17.4
H2	16.34	1.20	0.9965	15.08	1.02	0.9973	13.7
H3	21.14	0.99	0.9983	19.81	0.90	0.9971	14.5

Note: The index variables 'vol' and 'NMR' represent the results from the volumetric isotherm adsorption method and NMR isotherm adsorption method, respectively.

Figure 8 displays the relationships between $V_{L\text{-vol}}$ and coal's basic petrophysical parameters ($R_{o,m}$, vitrinite + inertinite content, FC_d, and A_d content). The $V_{L\text{-vol}}$ shows a 'U-morph' relationship, with $R_{o,m}$, having a minimum value at approximately ~1.6% (Figure 8a). The 'vitrinite + inertinite' content shows a positive relationship with the values of $V_{L\text{-vol}}$ (Figure 8b), but no significant correlations were found between the vitrinite content, inertinite content, and $V_{L\text{-vol}}$ (data not shown here). As shown in Figure 8c, the $V_{L\text{-vol}}$ is positively correlated with the FC_d content. Compared with the results in Figure 8a, the influence of FC_d content on Langmuir volume is weaker than that of $R_{o,m}$. Figure 8d displays the relationship between the A_d content and $V_{L\text{-vol}}$. The results show that, with the increase in the A_d content, $V_{L\text{-vol}}$ shows a decreasing trend, probably because the presence of ash can decrease the concentration of organic matter and reduce the methane adsorption capacity in coals [41].

3.3.2. NMR-Based Method

As discussed in Section 3.2, Equation (3) may lead to an erroneous adsorbed methane content. In the process of methane adsorption, only injected methane initially consists of a pressure-tight sample cell. Based on conservation of mass, the increase in adsorbed methane must have resulted entirely from the decrement in free methane. Similarly, the amount of adsorbed methane under specific pressures can be indirectly determined as:

$$V_{ads} = V_{tot} - V_{pip} - V_{fre} \tag{6}$$

where V_{ads} means the adsorbed methane content, cm^3; V_{pip} means the free methane content in the pipeline between two cells, 34.5 cm^3; V_{tot} means the total methane content in the sample cell, which can be determined according to the equation of state of ideal gas, cm^3; V_{fre} means the free methane content in sample cell, cm^3.

Similar to the calculation steps of the volumetric-based Langmuir parameter, the NMR-based adsorption experimental data, including adsorption pressure (P) and adsorption volume (V), are fitted by the Langmuir equation using Excel. As shown in Figure 7, the adsorption isotherms determined by the NMR method fitted well with the Langmuir equation.

3.3.3. Validity Application of the NMR Method

To verify the NMR method for quantitative evaluation of adsorbed phase methane content in coals, the results from the NMR method were compared with those from the volumetric-based method. Figure 7 shows the adsorption isotherms evaluated by the two different methods (volumetric- and NMR-based methods). The results show that the adsorption isotherms from these two methods have an excellent agreement for the selected coals (except for M2). Additionally, Figure 9 shows the scatter plots of the Langmuir volume obtained from the two methods contrasted in this study. The results show that all data points are approximately distributed on the 45° diagonal line (the dashed line in Figure 9), with the exception of sample M2. It is worth noting that the $V_{L\text{-NMR}}$ values are slightly less than $V_{L\text{-vol}}$ (Figure 9); this is probably because: (1) the minimum T_E value of NMR device in this study is 0.24 ms, thus does not detect the NMR distributions when T_2 is smaller than 0.24 ms; and (2) due to the methane adsorption exothermic in coals, the temperature error would have little effect on the experimental results. The relative deviation between the NMR and volumetric methods is ~5.95% (except for sample M2), which is within an allowable error. Additionally, the coal powders used in this study may result in a discrepancy in the experimental data for granulated samples because the grinding of samples would destroy the primary texture of the coal pore systems.

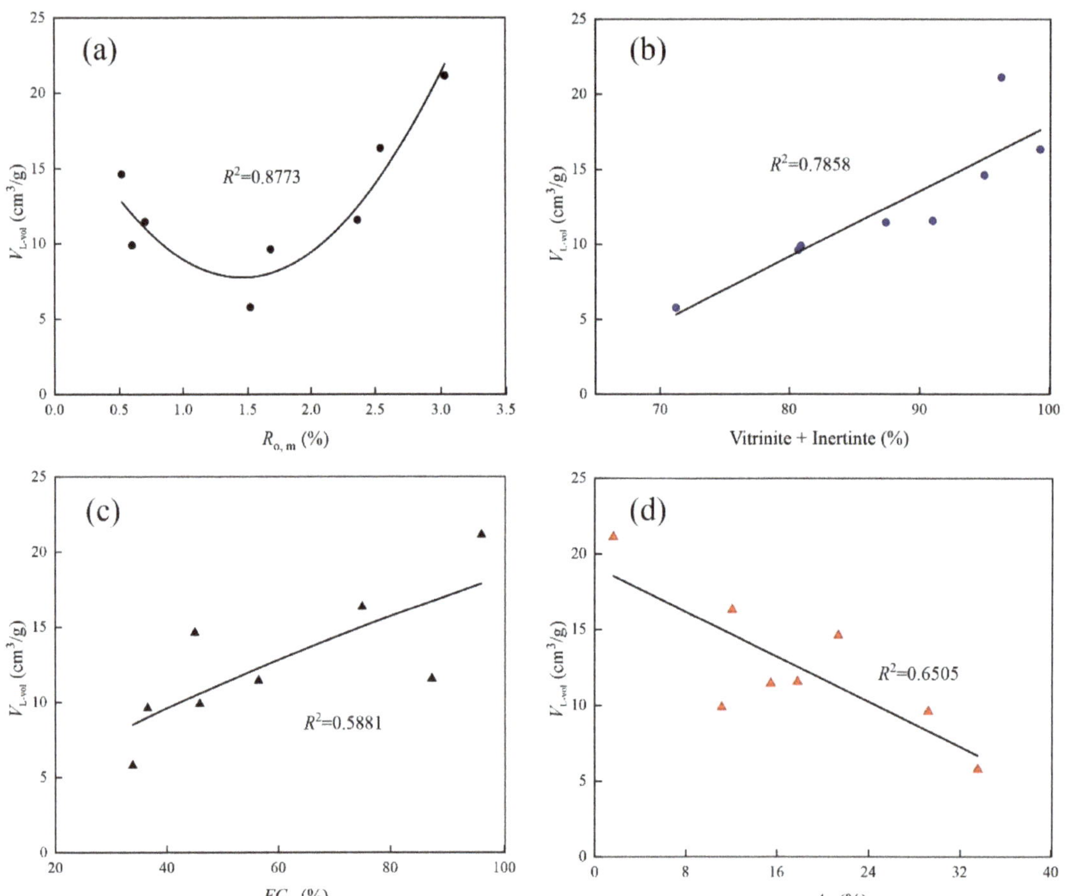

Figure 8. The cross-plots of $R_{o,m}$, vitrinite + inertinite content, FC_d content, and A_d content versus Langmuir volume $V_{L\text{-vol}}$. (**a**) $R_{o,m}$ versus Langmuir volume $V_{L\text{-vol}}$, (**b**) vitrinite + inertinite content versus Langmuir volume $V_{L\text{-vol}}$, (**c**) FC_d content versus Langmuir volume $V_{L\text{-vol}}$, (**d**) A_d content versus Langmuir volume $V_{L\text{-vol}}$.

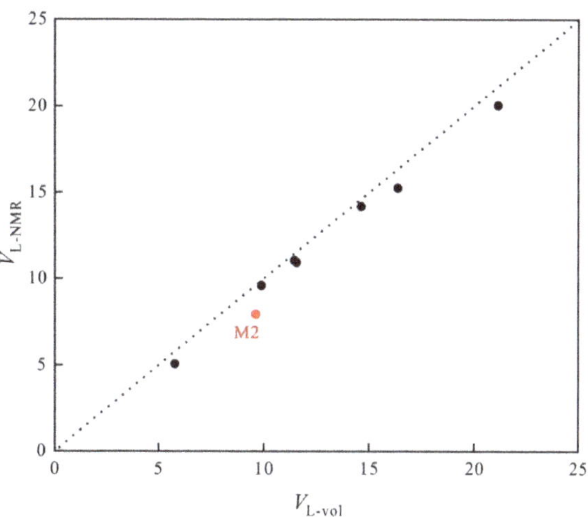

Figure 9. Scatter plots of Langmuir volume obtained from two contrasting methods.

For the sample M2, the adsorption isotherms determined from the NMR method show a large deviation from those of the volumetric-based method (Figure 7), and the Langmuir volume relative deviation is relatively high, at ~17.15%. Previous literature has found that paramagnetic minerals can form larger internal magnetic fields, and thus influence the NMR signal response [42]. SEM-EDS results demonstrate that some paramagnetic minerals were present in sample M2, namely, pyrite (Figure 10a) and ferrocalcite (Figure 10b). This probably results in a larger variation in methane adsorption capacity. The experimental results indicate that NMR T_2 measurements can be considered to be an effective method for the characterization of adsorbed methane content, even though some limitations exist due to paramagnetic minerals containing coals. Thus, the results observed in this study identify the directions for future work because the influence of paramagnetic minerals on NMR signals cannot be neglected.

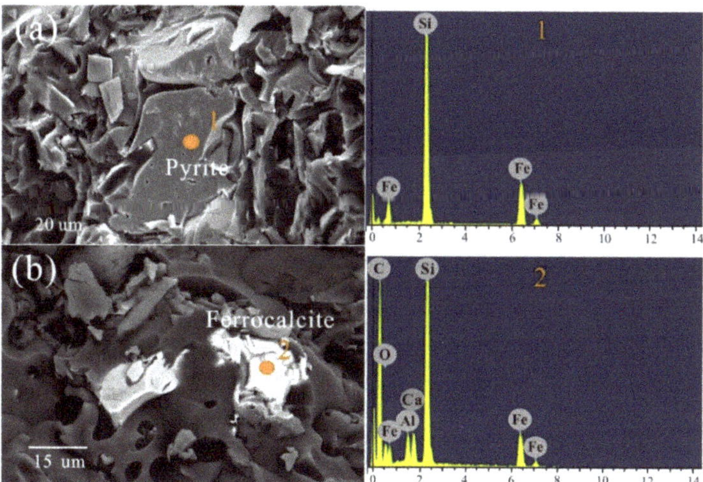

Figure 10. SEM-EDS results in sample M2. (**a**) pyrite minerals occured in sample M2, (**b**) ferrocalcite minerals occured in sample M2.

3.4. Methane Adsorption Dynamic Process Characterization

It is known that the Langmuir isotherm curves are only cross-plots of the adsorbed methane content and methane pressure. The shortcoming of this is that it is incapable of depicting any details of the methane adsorption dynamic process. An evident advantage of the NMR measurement designed in this study is that it can dynamically monitor multiphase methane in the methane adsorption process, because the NMR device can automatically measure the multiphase methane T_2 characteristics at any time. Here, the sample L2 is taken as a representative sample to discuss the application of the NMR relaxation method to dynamic characterization methane adsorption.

Figure 11a shows the NMR methane adsorption spectra under a methane pressure of 3.46 MPa with an interval time of 45 min. At the methane first injection, the NMR P1 peak shows a clear increasing trend, whereas the NMR P2 peak shows a decreasing trend. By comparison, the NMR P3 peak has changed to an obvious spectral peak. This is probably because of the larger amount of methane injected into the sample cell free space, which results in the clearer T_2 relaxation. The amplitude of the P1 peak increases rapidly from 1150.71 to 1652.53 during the first 45 min, and then decreases to an essentially stable level after 180 min (Figure 11b). The NMR P2 peak shows a rapidly decreasing trend during the first 135 min, and then increases slowly to reach an equilibrium condition.

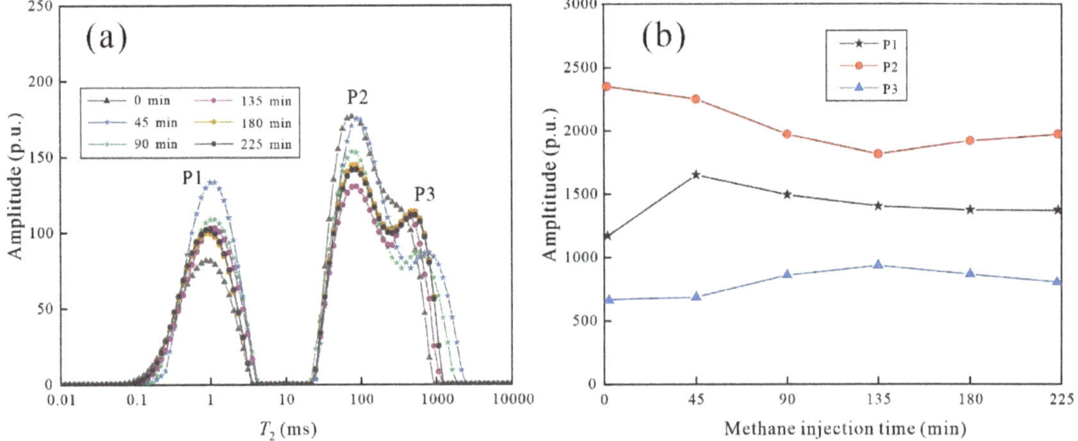

Figure 11. NMR T_2 spectra of methane adsorbed (**a**) and peak amplitude change characteristics (**b**) under a pressure of 3.46 MPa with an interval time of 45 min for sample L2.

Figure 12 displays the relationship between the adsorbed methane T_2 amplitude and injection time under six different pressures for sample L2. The results indicate that the NMR T_2 amplitude of the adsorbed methane increased rapidly at about 45 min, and then decreased to an equilibrium condition at approximately 3 h, indicating the adsorption time was approximately ~3 h. Compared with the conventional method (e.g., volumetric- or gravimetric-based methods), the unique features of the NMR-based method not only enable the determination of methane adsorption capacity, but also allow dynamic monitoring of multiphase methane in different experimental situations, such as methane adsorption/desorption and CO_2-enhanced CBM.

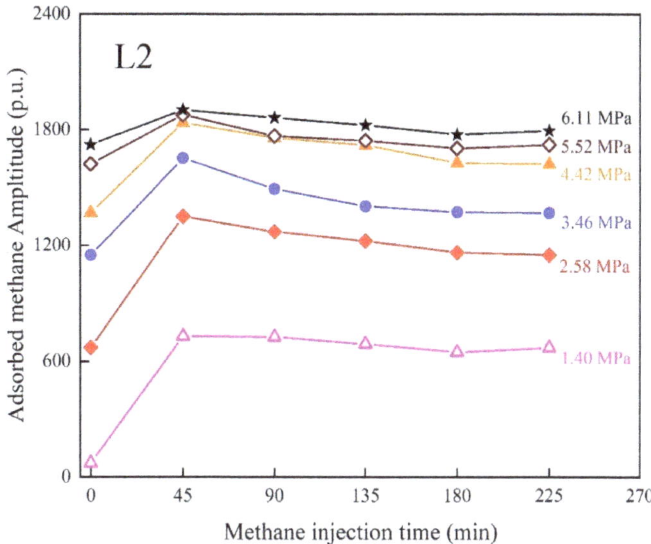

Figure 12. Relationship between adsorbed methane T_2 amplitude and injection time under six different pressures, for sample L2.

4. Conclusions

This paper proposed a novel method for the evaluation of the methane adsorption capacity and the quantitative characterization of the methane adsorption dynamic process based on NMR T_2 measurements. **The main conclusions are:**

(1) The NMR T_2 distributions of methane in coals exhibit three peaks, which are P1 ($T_2 < 8$ ms), P2 ($T_2 = 20$–300), and P3 ($T_2 > 300$ ms), corresponding to adsorbed phase methane, free phase methane between pores, and free phase methane in the sample cell, respectively.

(2) According to the results from the self-designed NMR isotherm adsorption measurements for eight coal samples, the NMR-based Langmuir volume ranges from 5.07 to 19.81 cm^3/g. The adsorption isothermal curves evaluated from the NMR and volumetric methods show an excellent agreement, and the Langmuir volume relative deviation, of approximately ~6.0%, was within an allowable limit.

(3) The NMR technique provides an alternative method for dynamic monitoring of multi-phase methane in the methane adsorption process, which is difficult to implement using conventional methods, such as the volumetric- and gravimetric-based methods

Author Contributions: S.S. and S.L. (Shiqi Liu) provided the funding acquisition; S.Z. performed the experiments and wrote the paper; M.W. and S.L. (Shijian Lu) analyzed the data; G.F. and J.H. revised the paper; Y.Y. and X.J. provided technical support. All authors have read and agreed to the published version of the manuscript.

Funding: This research was funded by the National Natural Science Foundation of China (grant number: 42141012; 42030810; 41972168), and A Project Funded by the Priority Academic Program Development of Jiangsu Higher Education Institutions.

Institutional Review Board Statement: Not applicable.

Informed Consent Statement: Not applicable.

Data Availability Statement: The data presented in this study are available on request from the corresponding author.

Conflicts of Interest: The authors declare no conflict of interest.

References

1. Fu, X.H.; Qin, Y.; Wang, G.X.; Rudolph, V. Evaluation of coal structure and permeability with the aid of geophysical logging technology. *Fuel* **2009**, *88*, 2278–2285. [CrossRef]
2. Karacan, C.Ö.; Ruiz, F.A.; Cotè, M.; Phipps, S. Coal mine methane: A review of capture and utilization practices with benefits to mining safety and to greenhouse gas reduction. *Int. J. Coal Geol.* **2011**, *86*, 121–156. [CrossRef]
3. Moore, T.A. Coalbed methane: A review. *Int. J. Coal Geol.* **2012**, *101*, 36–81. [CrossRef]
4. Cai, Y.D.; Liu, D.M.; Mathews, J.P.; Pan, Z.J.; Elsworth, D.; Yao, Y.B.; Li, J.Q.; Guo, X.Q. Permeability evolution in fractured coal—Combining triaxial confinement with X-ray computed tomography, acoustic emission and ultrasonic techniques. *Int. J. Coal Geol.* **2014**, *122*, 91–104. [CrossRef]
5. Li, Y.; Zhang, C.; Tang, D.Z.; Gan, Q.; Niu, X.L.; Wang, K.; Shen, R.Y. Coal pore size distributions controlled by the coalification process: An experimental study of coals from the Junggar, Ordos and Qinshui basins in China. *Fuel* **2017**, *206*, 352–363. [CrossRef]
6. Li, Y.; Tang, D.Z.; Wu, P.; Niu, X.L.; Wang, K.; Qian, P.; Wang, Z.S. Continuous unconventional natural gas accumulations of Carboniferous-Permian coal-bearing strata in the Linxing area, northeastern Ordos basin, China. *J. Nat. Gas Sci. Eng.* **2016**, *36*, 317–324. [CrossRef]
7. Cai, Y.D.; Liu, D.M.; Yao, Y.B.; Li, J.Q.; Qiu, Y.K. Geological controls on prediction of coalbed methane of No. 3 coal seam in Southern Qinshui Basin, North China. *Int. J. Coal Geol.* **2011**, *88*, 101–112. [CrossRef]
8. Zheng, S.J.; Yao, Y.B.; Liu, D.M.; Cai, Y.D.; Liu, Y.; Li, X.W. Nuclear magnetic resonance T_2 cutos of coals: A novel method by multifractal analysis. *Fuel* **2019**, *241*, 715–724. [CrossRef]
9. Li, J.Q.; Lu, S.F.; Xie, L.J.; Zhang, J.; Xue, H.T.; Zhang, P.F.; Tian, S.S. Modeling of hydrocarbon adsorption on continental oil shale: A case study on n-alkane. *Fuel* **2021**, *206*, 603–613. [CrossRef]
10. Li, J.Q.; Wang, S.Y.; Lu, S.F.; Zhang, P.F.; Cai, J.C.; Zhao, J.H.; Li, W.B. Microdistribution and mobility of water in gas shale: A theoretical and experimental study. *Mar. Petrol. Geol.* **2019**, *102*, 496–507. [CrossRef]
11. Hou, H.H.; Shao, L.Y.; Li, Y.H.; Li, Z.; Wang, S.; Zhang, W.L.; Wang, X.T. Influence of coal petrology on methane adsorption capacity of the Middle Jurassic coal in the Yuqia Coalfield, northern Qaidam Basin, China. *J. Pet. Sci. Eng.* **2017**, *149*, 218–227. [CrossRef]
12. Li, J.Q.; Lu, S.F.; Zhang, P.F.; Cai, J.C.; Li, W.B.; Wang, S.Y.; Feng, W.J. Estimation of gas-in-place content in coal and shale reservoirs: A process analysis method and its preliminary application. *Fuel* **2020**, *259*, 116266. [CrossRef]
13. Song, W.H.; Jun, Y.; Wang, D.Y.; Li, Y.; Sun, H.; Yang, Y.F. Dynamic pore network modelling of real gas transport in shale nanopore structure. *J. Pet. Sci. Eng.* **2020**, *184*, 105506. [CrossRef]
14. Pajdak, A.; Kudasik, M.; Skoczylas, N.; Wierzbicki, M.; Teixeira Palla Braga, L. Studies on the competitive sorption of CO_2 and CH_4 on hard coal. *Int. J. Greenh. Gas Con.* **2019**, *90*, 102798. [CrossRef]
15. Kudasik, M.; Skoczylas, N.; Pajdak, A. The repeatability of sorption processes occurring in the coal-methane system during multiple measurement series. *Energies* **2017**, *10*, 661. [CrossRef]
16. Genserblum, Y.; van Hemert, P.; Billemont, P.; Bush, A.; Charriére, D.; Li, D.; Krooss, B.M.; de Weireld, G.; Prinz, D.; Wolf, H.A.A. European inter-laboratory comparison of high-pressure CO_2 sorption isotherms. I: Activated carbon. *Carbon* **2009**, *47*, 2958–2969.
17. Ye, Z.H.; Chen, D.; Pan, Z.J.; Zhang, G.Q.; Xia, Y.; Ding, X. An improved Langmuir model for evaluating methane adsorption capacity in shale under various pressures and temperatures. *J. Nat. Gas Sci. Eng.* **2016**, *31*, 658–680. [CrossRef]
18. Krooss, B.M.; van Bergen, F.; Gensterblum, Y.; Siemons, N.; Pagnier, H.J.M.; David, P. High-pressure methane and carbon dioxide adsorption on dry and moisture-equilibrated Pennsylvanian coals. *Int. J. Coal Geol.* **2002**, *51*, 69–92. [CrossRef]
19. Coates, G.R.; Xiao, L.Z.; Primmer, M.G. *NMR Logging Principles and Applications*; Gulf Publishing Company: Houston, TX, USA, 1999.
20. Akkurt, R.; Vinegar, H.J.; Tutunjian, P.N.; Guillory, A.J. NMR logging of natural gas reservoirs. *Log. Anal.* **1996**, *37*, 33–42.
21. Howard, J.J. Quantitative estimates of porous media wettability from proton NMR. *Magn. Reson. Imaging* **1998**, *16*, 529–533. [CrossRef]
22. Zheng, S.J.; Yao, Y.B.; Liu, D.M.; Cai, Y.D.; Liu, Y. Characterizations of full-scale pore size distribution, porosity and permeability of coals: A novel methodology by nuclear magnetic resonance and fractal analysis theory. *Int. J. Coal Geol.* **2018**, *196*, 148–158. [CrossRef]
23. Zheng, S.J.; Yao, Y.B.; Liu, D.M.; Cai, Y.D.; Liu, Y. Nuclear magnetic resonance surface relaxivity of coals. *Int. J. Coal Geol.* **2019**, *205*, 1–13. [CrossRef]
24. Hari, P.; Chang, C.T.P.; Kulkarni, R.; Lien, J.R.; Watson, A.T. NMR characterization of hydrocarbon gas in porous media. *Magn. Reson. Imaging* **1998**, *16*, 545–547. [CrossRef]
25. Safiullin, K.; Kuzmin, V.; Bogaychuk, A.; Alakshin, E.; González, L.M.; Kondratyeva, E.; Dolgorukov, G.; Gafurov, M.; Klochkov, A.; Tagirov, M. Determination of pores properties in rocks by means of helium-3 NMR: A case study of oil-bearing arkosic conglomerate from North belt of crude oil, Republic of Cuba. *J. Petrol. Sci. Eng.* **2022**, *210*, 110010. [CrossRef]
26. Guo, R.; Mannhardt, D.; Kantzas, A. Characterizing moisture and gas content of coal by low-field NMR. *J. Can. Pet. Technol.* **2007**, *46*, 49–54. [CrossRef]
27. Vasilenko, T.A.; Kirillov, A.K.; Molchanov, A.N.; Pronskii, E.A. An NMR study of the ratio between free and sorbed methane in the pores of fossil coals. *Solid Fuel Chem.* **2018**, *52*, 361–369. [CrossRef]

28. Liu, J.; Yao, Y.B.; Liu, D.M.; Elsworth, D. Experimental evaluation of CO_2 enhanced recovery of adsorbed-gas from shale. *Int. J. Coal Geol.* **2017**, *179*, 211–218. [CrossRef]
29. Yao, Y.B.; Liu, J.; Liu, D.M.; Chen, J.Y.; Pan, Z.J. A new application of NMR in characterization of multiphase methane and adsorption capacity of shale. *Int. J. Coal Geol.* **2019**, *131*, 76–85. [CrossRef]
30. Yao, Y.B.; Liu, D.M.; Xie, S.B. Quantitative characterization of methane adsorption on coal using a low-field NMR relaxation method. *Int. J. Coal Geol.* **2014**, *201*, 32–40. [CrossRef]
31. Li, Y.; Wang, Y.B.; Wang, J.; Pan, Z.J. Variation in permeability during CO_2–CH_4 displacement in coal seams: Part 1—Experimental insights. *Fuel* **2020**, *263*, 116666. [CrossRef]
32. Yao, Y.B.; Liu, D.M.; Che, Y.; Tang, D.Z.; Tang, S.H.; Huang, W.H. Petrophysical characterization of coals by low-field nuclear magnetic resonance (NMR). *Fuel* **2010**, *89*, 1371–1380. [CrossRef]
33. Yuan, Y.J.; Rezaee, R.; Verrall, M.; Hu, S.Y.; Zou, J.; Testmanti, N. Pore characterization and clay bound water assessment in shale with a combination of NMR and low-pressure nitrogen gas adsorption. *Int. J. Coal Geol.* **2018**, *194*, 11–21. [CrossRef]
34. Zhao, Y.X.; Zhu, G.P.; Dong, Y.H.; Danesh, N.N.; Chen, Z.W.; Zhang, T. Comparison of low-field NMR and microfocus X-ray computed tomography in fractal characterization of pores in artificial cores. *Fuel* **2017**, *210*, 217–226. [CrossRef]
35. Sun, X.X.; Yao, Y.B.; Liu, D.M.; Zhou, Y. Investigations of CO_2-water wettability of coal: NMR relaxation method. *Int. J. Coal Geol.* **2018**, *188*, 38–50. [CrossRef]
36. Zheng, S.J.; Yao, Y.B.; Liu, D.M.; Cai, Y.D.; Liu, Y. Quantitative characterization of multiphase methane in coals using the NMR relaxation method. *J. Petrol. Sci. Eng.* **2021**, *198*, 108148. [CrossRef]
37. Perera, M.S.A.; Ranjith, P.G.; Choi, S.K.; Airey, D.; Weniger, P. Estimation of gas adsorption capacity in coal: A review and an analytical study. *Int. J. Coal Prep. Util.* **2012**, *32*, 25–55. [CrossRef]
38. Siemons, N.; Wolf, K.A.A.; Bruining, J. Interpretation of carbon dioxide diffusion behavior in coals. *Int. J. Coal Geol.* **2007**, *72*, 315–324. [CrossRef]
39. Weniger, P.; Franců, J.; Hemza, P.; Krooss, B.M. Investigations on the methane and carbon dioxide sorption capacity of coals from the SW upper Silesian Coal Basin Czech Republic. *Int. J. Coal Geol.* **2012**, *93*, 23–39. [CrossRef]
40. Zhang, M.; Fu, X.H. Characterization of pore structure and its impact on methane adsorption capacity for semi-anthracite in Shizhuangnan Block, Qinshui Basin. *J. Nat. Gas Sci. Eng.* **2018**, *60*, 49–62. [CrossRef]
41. Laxminarayana, C.; Crosdale, P.J. Role of coal type and rank on methane sorption of Bowen Basin, Australia coals. *Int. J. Coal Geol.* **1999**, *40*, 309–325. [CrossRef]
42. Sadian, M.; Prasad, M. Effect of mineralogy on nuclear magnetic resonance surface relaxivity: A case study of Middle Bakken and Three Forks formations. *Fuel* **2015**, *161*, 197–206. [CrossRef]

Article

Genesis of Bedding Fractures in Ordovician to Silurian Marine Shale in Sichuan Basin

Hu Wang [1,2], Zhiliang He [3,4,*], Shu Jiang [1,2,*], Yonggui Zhang [3,4], Haikuan Nie [3,4], Hanyong Bao [5] and Yuanping Li [1,2]

[1] Key Laboratory of Tectonics and Petroleum Resources, Ministry of Education, China University of Geosciences, Wuhan 430074, China
[2] School of Earth Resources, China University of Geosciences, Wuhan 430074, China
[3] China State Key Laboratory of Shale Oil and Gas Enrichment Mechanisms and Effective Development, Beijing 100083, China
[4] Sinopec Petroleum Exploration and Development Research Institute, Beijing 100083, China
[5] Exploration and Development Research Institute, SINOPEC Jianghan Oilfield Company, Wuhan 430073, China
* Correspondence: wanghu0622@cug.edu.cn (Z.H.); jiangsu@cug.edu.cn (S.J.)

Citation: Wang, H.; He, Z.; Jiang, S.; Zhang, Y.; Nie, H.; Bao, H.; Li, Y. Genesis of Bedding Fractures in Ordovician to Silurian Marine Shale in Sichuan Basin. *Energies* 2022, 15, 7738. https://doi.org/10.3390/en15207738

Academic Editors: Manoj Khandelwal and Ingo Pecher

Received: 20 June 2022
Accepted: 10 October 2022
Published: 19 October 2022

Abstract: The effective utilization of shale bedding fractures is of great significance to improve shale gas recovery efficiency. Taking the Wufeng–Longmaxi Formation shale in Sichuan Basin as the research object, the formation process and mechanism of bedding fractures in marine shale are discussed, based on field observation and description, high-resolution electron microscope scanning, fluid inclusion detection, and structural subsidence history analysis. The results show that the formation of bedding fractures is jointly controlled by sedimentary characteristics, hydrocarbon generation, and tectonic movement: the development degree of bedding (fractures) is controlled by the content of shale organic matter and brittle minerals, and bedding fractures formed in the layers with high organic matter; tectonic movement created stress environment and space for bedding fractures and promoted the opening of bedding fractures; the time for calcite vein to capture fluid is consistent with the time of oil-gas secondary pyrolysis stage. The formation of the calcite vein is accompanied by the opening of fractures. The acid and oil-gas generated in the hydrocarbon generation process occupied the opening space and maintained the bedding fractures open. The study of the formation process of bedding fractures is helpful to select a suitable method to identify bedding fractures, and then effectively use it to form complex fracture networks in the fracturing process to improve shale oil and gas recovery.

Keywords: fractures; complex fracture net; formation of bedding; Fuling area

1. Introduction

At present, shale gas resources are mainly found in Sichuan Basin and Ordos Basin, with an estimated resource volume of 80 trillion m^3. By December 2020, shale gas production has reached 20.04 billion m^3; continental shale oil is mainly concentrated in northern China. According to the public data of major oil companies, including 11 basins such as Songliao, Bohai Bay, Junggar, and Ordos, the shale oil resources with medium and high maturity (Ro: 0.7–1.3) are 10 billion tons. It is estimated that by the end of the "15th Five-Year Plan", the annual output may reach 40 million tons, which will account for 20% of total crude oil production [1].

The production of shale oil and gas is affected by geological and engineering factors. Geological factors are mainly organic matter abundance, the thickness of the sweet section, formation pressure, clay mineral content, fracture development intensity, bedding density, and other factors while engineering factors are mainly the horizontal well length and the number of fracturing section clusters spacing and casing deformation [2–4]. Previous research results believe that bedding controls the distribution of organic matter and brittle minerals to a certain extent, and bedding fractures are important reservoir spaces of the Wufeng–Longmaxi formation shale in the Sichuan Basin [5]. The formation of bedding fractures is generally closely related to the stress field, and the bedding fractures are in multiple scales [6,7]. The late uplift and reconstruction of the Sichuan Basin are of great significance to the formation and activation of fractures [8]. The differences in development intensity and longitudinal stress in weak bedding planes will directly affect the development scale and complexity of fractures in the reservoir after fracturing [9]. The study on shale optimization schemes in Permian basins in North America found that the cluster spacing was shortened to 6.1 m, the complex fracture network produced the largest volume, and the recoverable reserves could be increased by 140% [10]. Generally, the bedding fractures develop in small fracture height, but they will increase the spatial complexity of hydraulic fractures, which is conducive to improving the production and ultimate recovery of a single well [11].

However, most of these studies on bedding fractures are based on the effects and results brought by bedding fractures. There are few studies on the controlling factors of the formation of bedding fractures, which is an urgent need for research on bedding fractures to provide theoretical support for the identification and prediction of bedding fractures. Based on the data of cores and outcrops, this paper attempts to analyse the controlling factors of shale bedding fracture formation by means of rock mineral analysis, scanning electron microscope, and fluid inclusion analysis, so as to provide a geological basis for identifying and characterizing the three-dimensional spatial distribution of bedding fracture, optimize the fracturing network reconstruction scheme, and improve shale oil and gas recovery.

2. Geological Background and Data Method

2.1. Geological Background

The eastern edge and southern edge of the Sichuan Basin mainly include Chongqing and Sichuan Province, separately. The southeast edge of the basin formed a strongly rising fold belt with a high altitude, where are middle and low mountain areas. The villages and towns in the area are connected by simple roads with inconvenient traffic. The study area is located in the east of the Sichuan Basin, northeast of the Fuling area, Chongqing (Figure 1a). Affected by the central Sichuan uplift, the central Guizhou paleouplift, and the Jiangnan Xuefeng mountain paleouplift, the shale reservoirs of the Wufeng–Longmaxi Formation are mainly distributed in the southeast of the Sichuan Basin and its periphery. In the Chongqing Nanchuan, the shale of the Longmaxi Formation is missing due to the influence of the underwater highlands in the northeast of the central Sichuan paleouplift and the northeast of the central Guizhou paleouplift. The overall thickness of high-quality shale varies from 5 m to 40 m, and the target layer in the study area is just in the area with the largest thickness (Figure 1). During the shale deposition period of the Wufeng Formation, the water was deep, which was also a high-quality section of the shale gas reservoir. The deposition thickness in the target area was 6 m, which was the center of the depression area. The high-quality reservoir section of the Wufeng–Longmaxi Formation was deposited in an oxygen-deficient or anoxic environment and rich in organic matter. However, in the upper Longmaxi Formation, the sedimentary environment changed, resulting in a decrease in organic matter.

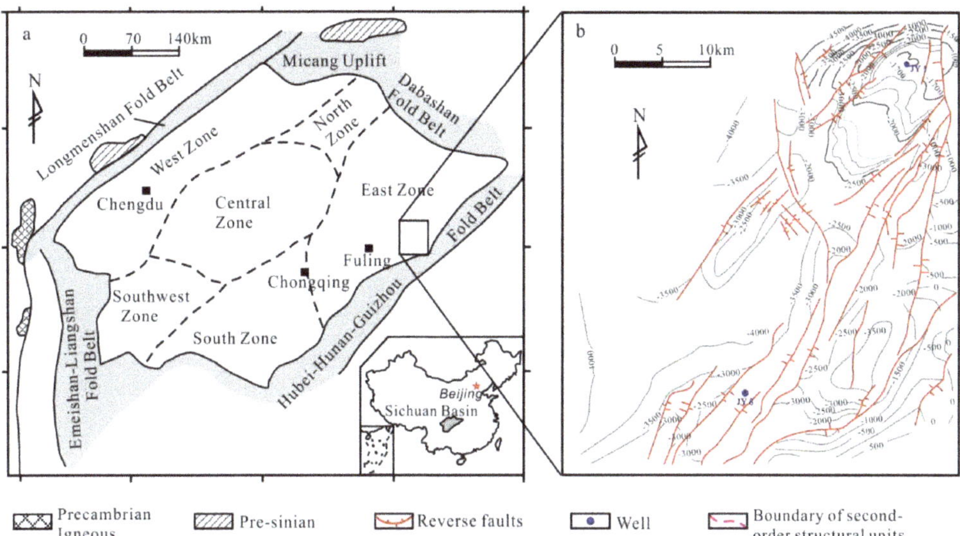

Figure 1. Location structure of study area. (**a**) Regional division of Sichuan Basin, (**b**) Structural plan of Jiaoshiba area.

2.2. Data and Experimental Method

2.2.1. Data

The core and outcrops samples are mainly collected from the Qijiang Guanyinqiao, Pengshui Lujiao, and Shizhu Qiliao sections and Jiaoshiba shale gas field in Chongqing at the Wufeng Formation of Upper Ordovician and Longmaxi Formation of Lower Silurian, such as Well JY1, Well JY11-4, Well JY41-5, Well Pengye 1, Well Yucan 4, Well Yucan 6, Well Yongye 6, etc. Quartz vein and calcite vein were mainly collected in coring wells, and the depth distribution range of samples is between 2250~2620 m. The 276 km^2 three-dimensional seismic data of the Wufeng–Longmaxi Formation in the Fuling area of the Sichuan Basin were sorted out and interpreted. All experiments were completed in the Key Laboratory of shale oil and gas exploration and development, Wuxi Institute of geology, Sinopec.

2.2.2. X-ray Diffraction

The mineral composition analysis of the samples is based on SY/T 5163-2018 "X-ray diffraction analysis method for clay minerals and common non-clay minerals in sedimentary rocks", using Rigaku Ultima IV X-ray diffractometer (Tokyo, Japan). The maximum rated voltage of the instrument is 60 mA, the maximum rated current is 40 mA, DOPS direct optical positioning system is adopted, and the angle reproducibility is better than ±0.0001°. The test is carried out in an environment of 25 °C, the working voltage is 20–40 Kv, and the scanning current is 10–40 mA. Before XRD analysis, the sample was crushed to 200 mesh.

2.2.3. Fluid Inclusion Analysis

In this study, isolated primary brine inclusions and gas-liquid two-phase brine inclusions associated with pure methane inclusions are selected for homogenization temperature test, so that the homogenization temperature of fluid inclusions can accurately reflect the maximum paleogeothermal temperature experienced by shale. The rock samples were made into a double-sided polished fluid inclusion sheet, and the micro temperature measurement of the inclusions in the vein was carried out by using the Axio scope A1 microscope of Zeiss company (Jena, Germany), equipped with the mdsg600 cold and hot platform of Linkam company (Redhill, UK) and the corresponding cold and hot systems. The temperature measurement accuracy of the hot and cold platforms is ± 0.1 °C. During the determination of the homogenization temperature of fluid inclusions, the initial heating rate is 15 °C/min, and when the inclusions are close to homogenization, it drops to 2 °C/min.

3. Results

3.1. Lithologic Profile

The characteristics of field outcrops can clearly show the abrupt change from the Guanyinqiao shell marl of the Wufeng Formation in Upper Ordovician to the organic rich shale at the bottom of the Silurian Longmaxi Formation, and then the Longmaxi shale gradually became organic poor shale upward. The field outcrop observation and rock mineral, geochemical, and petrophysical tests of the Lujiao section show that Wufeng is siliceous shale, and Guanyinqiao shell limestone cannot be found. The lower part of Longmaxi Formationis rich in organic matter shale, which transits upward to poor organic matter shale and silty shale. It can be seen that lithofacies, geochemistry, and petrophysical heterogeneity are relatively strong. The Qiliao section also has similar heterogeneity in a vertical direction. The Wufeng Formation transits from the lower siliceous shale to the upper marl interlayer and shell limestone. The lower part of Longmaxi shale is organic-rich carbonaceous shale deposited in deep-water shelf or depression on the shelf, and the upper part is deficient organic matter, gray matter, and sandy shale (Figure 2). All organic shale and siliceous shale contain abundant graptolites.

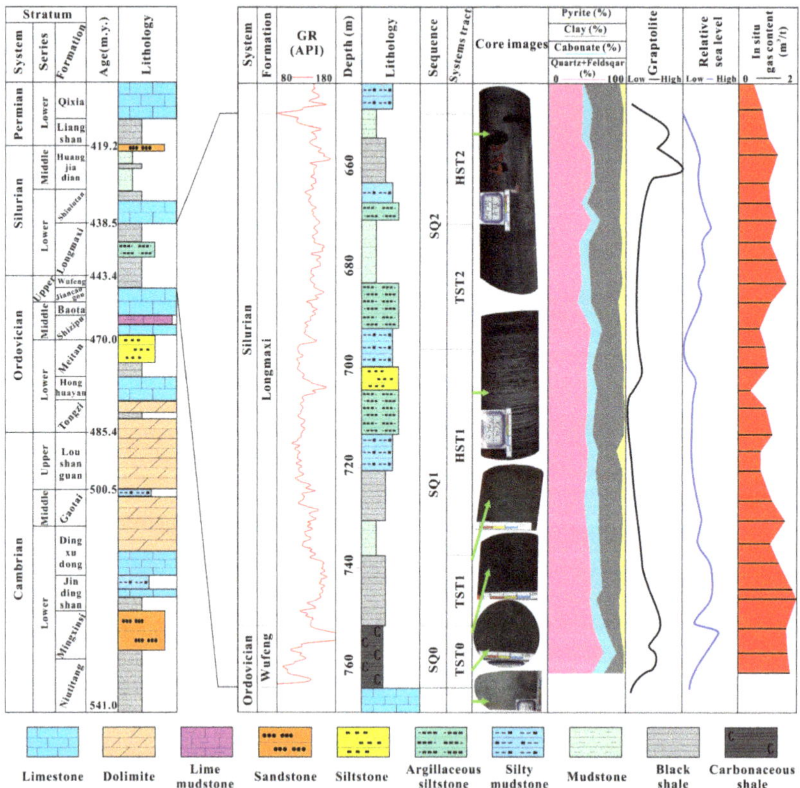

Figure 2. Sedimentary bottom framework of Wufeng–Longmaxi Formation in Sichuan Basin.

3.2. Bedding Characteristics

The main types of Longmaxi Formation shale are carbonaceous shale, calcareous shale, and silty shale, which respectively deposited in deep-water, relatively deep-water, and shallow-water environments. The thickness of the strata and the degree of bedding development vary greatly. The thickness of carbonaceous shale varies greatly from 0.1 mm to 10~30 cm. The thickness distribution of silty shale is relatively uniform, generally between 1~30 mm. Calcareous shale is less developed, and its thickness is generally less than 20 mm (Figure 3).

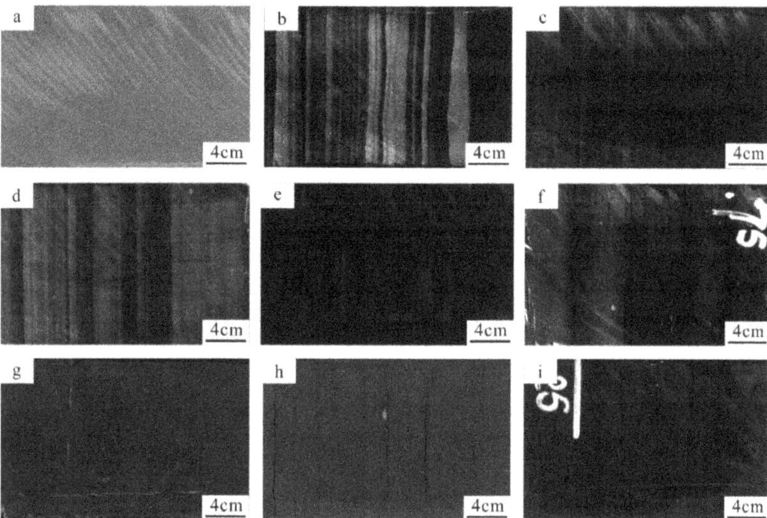

Figure 3. Bedding development of shale with different lithology. (**a**) Grey argillaceous siltstone; (**b**) Interbedding of grey black carbonaceous shale and grey argillaceous siltstone; (**c**) Gray black carbonaceous shale; (**d**) Gray black siliceous shale; (**e**) Black carbonaceous shale; (**f**) Gray black carbonaceous shale; (**g**) Gray black carbonaceous shale; (**h**) Gray black carbonaceous shale; (**i**) Gray black carbonaceous shale.

The linear density of bedding (fracture) of shale reservoir varies widely, with at least 127 layers/m (corresponding to 84 bedding fractures/m) and at most 1597 layers/m (corresponding to 480 bedding fractures/m). The quantity of fractures changes frequently. The linear density of bedding of 43 samples fluctuates up and down 14 times with the change of depth. The number of bedding at the bottom of the Wufeng Formation is the largest, and the overall development degree of bedding from the Wufeng Formation to the upper is gradually decreasing (Figure 4). The number of bedding at the bottom of the Wufeng Formation is the most, and the overall development degree of bedding from the Wufeng Formation to the upper is gradually reduced.

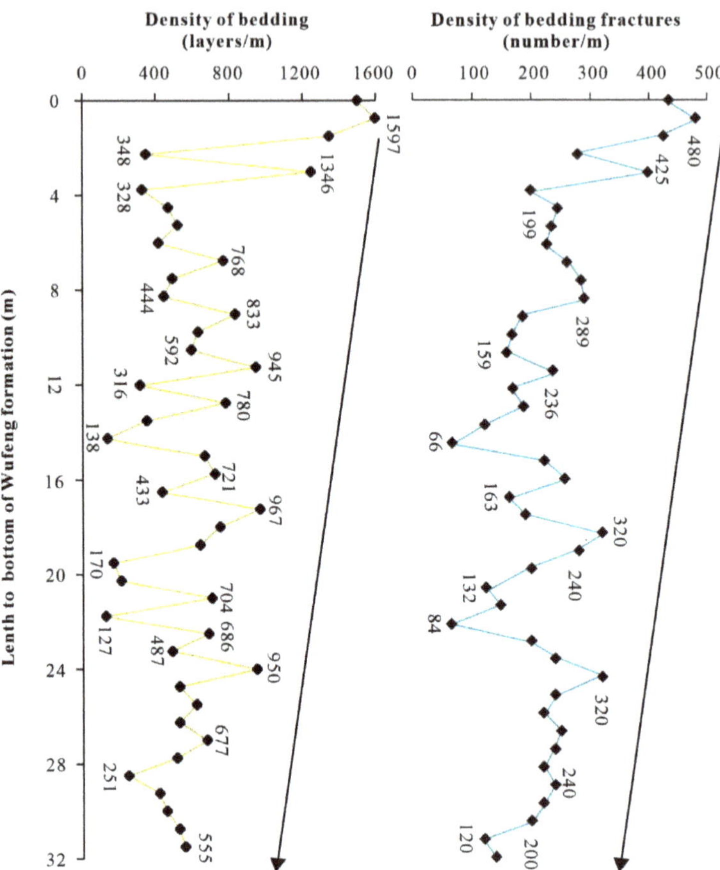

Figure 4. Longitudinal distribution of bedding and fracture density.

3.3. Stress Background

The strike of the main large faults in the Fuling gas field is basically in NE, but the faults in the north of Jiaoshiba are obviously biased to the left, and the small NW faults in the west of the internal Wujiang fault and Diaoshuiyan fault wre formed in the late stage. It is considered that the NE structure in the study area was formed in the early stage, and was compressed by the NW structure in the late stage to form the NW faults, which transformed the existing NE faults and formed the fault pattern in the current target study area. The study area mainly exists in two groups of northeast and northwest faults, which were formed in the late Yanshan movement and are the result of the East-West compressive stress caused by the subduction of the Pacific plate. The movement of the NE trending fault is larger than that of the NW trending fault. The movement of several large boundary faults is very large, with an extension distance of more than 10 km and a fault displacement of more than 300 m, which penetrated the Cambrian to Triassic strata (Figure 5).

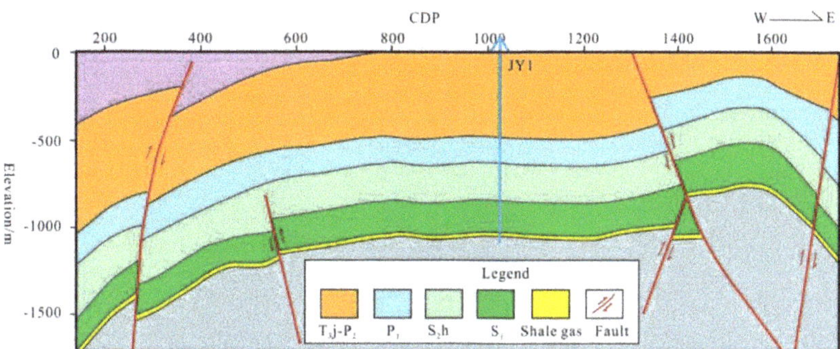

Figure 5. Characteristics of EW trending faults in Jiaoshiba block.

The Wufeng–Longmaxi Formation in the Fuling gas field experienced all the tectonic movements since the Caledonian movement. However, according to the analysis of the subsidence history of Well JY1, the whole target strata was in the stage of continuous decline of buried depth before the Hercynian movement, and there was no large tectonic uplift. Until the late Hercynian movement, the target strata in the whole study area began to rise rapidly and transform into the current structural pattern. It is believed that the Yanshan movement played a controlling role in the structure of the target strata in the study area. Further studies have found that Well JY1 experienced three periods of tectonic uplift: Late Yanshanian, late Yanshanian-early Himalayan, and early Himalayan to now. Only the tectonic uplift in late Yanshanian caused strong tectonic deformation in the study area. It is considered that the structural form of the Longmaxi Formation in Wufeng, Fuling District is mainly affected by the early and middle Yanshan tectonic movements. In the early Yanshan movement, the target area was affected by compressive stress, forming a syncline structural form. In the middle Yanshanian period, the target area experienced three deformation stages: basement thrust, cap rock detachment, and left lateral compression and torsion, forming the current anticline structure in the Fuling gas field.

3.4. Homogenization Temperature of Fluid Inclusions

Well JY1, JY4, JY5, JY7, JY8, and JY41-5 were selected for fracture inclusion analysis in the Fuling gas field (Figure 6). The results of micro temperature measurement and micro laser Raman spectroscopy show that the types of fluid inclusions captured in the veins are mainly gas-liquid two-phase brine inclusions, single-phase pure methane inclusions, hydrocarbon-containing brine inclusions, asphalt inclusions, etc.

According to the statistics of the homogenization temperature of fracture vein inclusions in six wells of the Fuling gas field, the homogenization temperature of gas-liquid two-phase inclusions in fractured calcite of Wufeng Formation of Well JY1 is distributed at 160~250 °C, concentrated at 160~170 °C and 200~230 °C (Figure 7). The homogenization temperature of calcite inclusions in the Wufeng Formation of Well JY4 is mainly distributed at 150~160 °C and 190~200 °C. The homogenization temperature of quartz inclusions in the upper Longmaxi Formation (2509.3 m) of Well JY41-5 is mainly 210~240 °C. The temperature measured by other wells in the adjacent area is in the range of 110 °C, 140~160 °C, and 180~200 °C, indicating the fluid activity period is more than that in Well JY1, JY4, and JY41-5 areas. Both organic and inorganic genetic mechanisms may generate calcite veins: atmospheric water, primary water, salt water, etc. Gao et al. [12] believed that calcite in the study area formed in a strong water-rock reaction. In the study area, the occurrence of calcite veins has a positive correlation with TOC content in the target interval, and the existence of hydrocarbon inclusions in calcite veins also indicates that calcite veins have a direct relationship with organic matter.

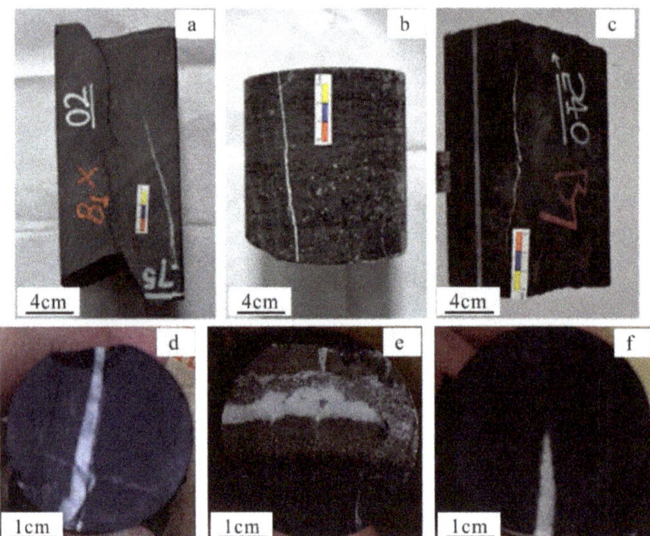

Figure 6. Fracture-filled calcite vein shale sample. (**a**) Longmaxi Formation shale, structural fracture, calcite vein filling; (**b**) Longmaxi Formation shale, structural fracture, and quartz vein filling; (**c**) Longmaxi Formation shale, structural fracture, calcite vein filling; (**d**) Longmaxi Formation shale, calcite vein filling, non-structural fracture; (**e**) Longmaxi Formation shale, filled with pyrite and calcite veins, non-structural fractures; (**f**) Longmaxi Formation shale, calcite vein filling, non-structural fracture.

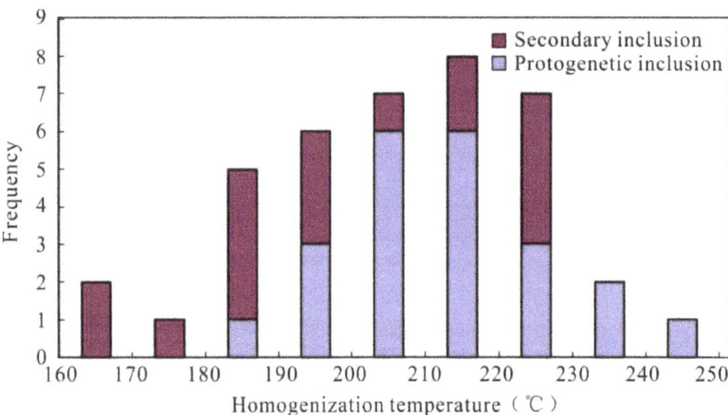

Figure 7. Homogenization temperature distribution histogram of fluid inclusions in calcite vein of Well JY1.

4. Discussion

The formation of laminae is due to the periodic or seasonal changes in physical or chemical conditions in a sedimentary environment [13–15]. There is little research on the formation of bedding fractures among beds. From the geological theory, it can be divided into internal and external causes: internal causes mainly include lithofacies and mineral composition characteristics; external factors mainly include regional tectonism, structural location, sedimentary diagenesis, and high abnormal pressure generated during hydrocarbon generation [16–18]. In fact, rock lithofacies and mineral structure are the internal manifestations of sedimentation. From the perspective of sedimentary phenomena, tectonic background, and hydrocarbon generation evolution process, this paper analyzes the controlling factors for the formation of foliation fractures in the Longmaxi Formation of Sichuan Basin, and puts forward an evolution model of "sedimentary bedding cast foundation, tectonic compression supply space, and hydrocarbon generation process fixation fractures".

4.1. Sedimentary Bedding Cast Foundation

The change in the sedimentary environment mainly affects the formation of non-structural fractures. For shale reservoirs, the change in the sedimentary environment directly affects the change in lithology. The density of lamina development in shale reservoirs with different lithology varies greatly, which directly determines the number of weak planes of shale reservoirs. It is also the basis for the development of bedding fractures. The bedding in the Longmaxi Formation shale of Jiaoshiba is well developed, which is formed by the interaction of silty, calcareous, and organic carbonaceous mudstone micro-layers, and the bedding thickness is different. Yue et al. [19] through the statistical work on the fracture density on the shale outcrop of the Niutitang Formation, believe that there is a negative correlation between the thickness of the shale reservoir and the fracture density. In addition, the research shows that different sedimentary environments will produce different types of clay minerals, which will have a direct impact on the fractures [20]. Due to the stability of marine sedimentation, the horizontal shale rock type, clay content and bedding development degree in the study area have basically not changed, but the changes in the overall environment of the basin in different periods have resulted in the changes in the bedding thickness, color, and mineral structure of each small layer in the vertical direction. Therefore, it can be inferred that the sedimentary environment is the key factor controlling the degree of bedding development, and the basic characteristics of the large development of organic matter, brittle minerals, and bedding in the Longmaxi Formation shale reservoir have cast a good foundation for the development of bedding fractures (Figure 3).

There is a good positive relationship between the number of cracks and the number of rhythmic segments. The rhythmic sections imply frequent changes in the sedimentary environment; the greater the difference between layers, the easier it is to form bedding contact surfaces with weak mechanical properties (Figure 4).

Through statistical analysis of rock slice and whole rock mineral characteristics of Well JY11-4 and well JY41-5, it is found that small layer 1–9 of the Wufeng–Longmaxi Formation in the Fuling gas field has small horizontal difference and large vertical difference, and the lithological difference is closely related to sea level fluctuation. From the bottom to the top of the study area, comparing layers 4 and 9 with relatively low Young's modulus and Poisson's ratio, the 1–3 and 5–8 layers with high Young's modulus and Poisson's ratio have high fracture development. This difference has a good correspondence with the change in sea level. Relatively high sea level generally deposits clay rock with high TOC and low brittle minerals, while small layer 1–9 shale has high TOC and brittle minerals, which was controlled by the change of sedimentary environment and diagenetic biogenesis. However, the strength of biogenic silica is consistent with that of all shale, that is, the change in sedimentary environment controls the lithological difference, while the lithological difference controls the difference in TOC and brittle mineral content, resulting in the difference in compressibility and compressive strength of small layers 1–9, which is an important internal factor controlling the development degree of fractures (Figure 8). The development degree of bedding fracture is controlled by the lithology of the shale reservoir. The development degree of micro-fracture in different shale lithology is different and regular. The content of organic matter in the shale reservoir also controls the development degree of shale microfracture. The development degree of microfracture is high in the parts with high content of organic matter.

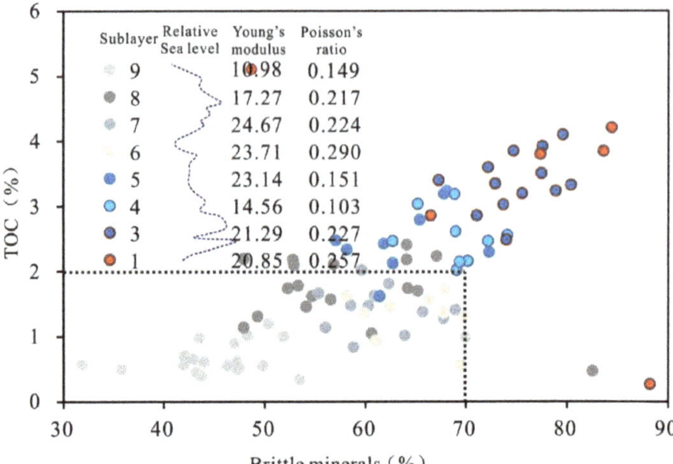

Figure 8. Relationship between TOC, brittle mineral content, and sedimentary environment.

4.2. Structural Extrusion Space

After the hydrocarbon generation peak of the Wufeng–Longmaxi Formation shale, it experienced three major tectonic movements: the Indosinian period, Yanshanian period, and Himalayan period, forming a complex structural style with thrust nappe, strike-slip, and tension multi-stress background. The faults and fractures generated by Indosinian and subsequent structural movements redistribute and readjust oil and gas accumulation. Tectonism is the most important factor affecting fractures, and the types of fractures formed under different tectonic backgrounds are different. It is easy to form low-angle slippage joints, interlayer slippage joints, and interlayer foliation joints under the background of extrusion stress. Tensile cracks are easy to form under the background of tensile stress. In addition to all the above cracks, tensile torsional and compressive torsional cracks can also occur under the strike-slip background [6].

The relationship between structural deformation and cracks was first applied to material mechanics. They believe that materials will be deformed during extrusion, in which cracks will be formed at the bending part of the deformation, and there will be a neutral plane in the middle of the material [21]. This is similar to the anticline, syncline, and fold structures formed by the compression of the strata. Different bending degrees have a direct linear relationship with the development intensity of fractures [22,23]. The thickness, Young's modulus, Poisson's ratio, and other parameters of Ordovician Cambrian strata in the Sichuan Basin are collected to calculate the position of the tectonic neutral plane of the strata in this area (Table 1). The results show that the structural neutral plane of the Fuling gas field in the Sichuan Basin is developed in the Cambrian strata at the bottom; that is, the Silurian Longmaxi Formation and Ordovician Wufeng Formation are both located in the upper part of the fold neutral plane, belonging to the compression area, which is easy to form low angle slip fractures, interlayer slip fractures, interlayer foliation fractures, and the changes of structural morphology are directly related to the genesis of fractures.

$$H = \frac{\sum\limits_{i=1}^{n} \frac{E_i}{1-\mu_i^2} \left| 2h - h_i - 2\sum\limits_{j=0}^{i-1} h_j \right|}{2\sum\limits_{i=1}^{n} \frac{E_i}{1-\mu_i^2} h_i} \tag{1}$$

where, h is the distance from the neutral surface of the fold to the stratum floor, m; H is the total thickness of formation, m; h_i is the thickness of different strata, m; E_i is Young's modulus of different formations, 10^4 pa; μ_i is Poisson's ratio of different strata dimensionless.

Table 1. Statistics of stratigraphic parameters in the southeast edge of Sichuan Basin.

Stratum	Group	Litho	Thickness (m)	Young's Modulus (10^4 Pa)	Poisson's Ratio
Silurian	Hanjiadian	Mudstone	317	3.131	0.23
	Xiaoheba	Siltstone, shale	160	2.658	0.15
	Longmaxi	Shale	488	3.037	0.21
Ordovician	Wufeng	Shale	6	2.449	0.13
	Linxiang	Limestone	20		
	Baota	Limestone	20–25	0.538	0.35
	Shizipu	Limestone	20		
	Dawan	Limestone	50		
	Tongzhi	Dolomite	100	4.522	0.25
Cambrian		Limestone	640	0.538	0.35

4.3. Solid Fracture during Hydrocarbon Generation

The evolution of hydrocarbon and fracture generation in shale reservoirs are well matched in time and space. Immature stage: before the burial depth reaches 1500 m for the first time, 50–60 °C, before the Middle Silurian, which is mainly the stage of organic matter conversion to kerogen and biogenic methane generation with biological activity. This stage is not directly related to the formation of fractures, but the active biological activities and a large number of biological quartz in this stage provide the material basis for the later generation of fractures [24,25]. Plutonic stage: burial depth is gradually from 1500 m to 6000 m, 60–200 °C, Middle Silurian to early Cretaceous, a large amount of organic matter transformed into liquid hydrocarbons, and organic acids precipitated at the same time, resulting in dissolution. After the organic acid is consumed in the later stage, iron calcite and iron dolomite are precipitated to fill the cracks and increase the effectiveness of the fractures [26]. In the middle diagenetic stage, the formation of dissolution pores and fractures by dissolution and the filling of fractures by recrystallization increase the effectiveness of fractures. Metamorphism stage: after the burial depth of 6000 m, after the early Cretaceous, montmorillonite has been completely transformed into illite, hydrocarbon evolution has reached an over-mature stage, and the secondary cracking of oil and asphalt forms dry gas. The formed dry gas forms fluid overpressure, which is conducive to the formation of fractures.

Gao [12,27] et al. believed that through isotope measurement and fluid inclusion analysis, the formation time of calcite vein in the Longmaxi Formation shale reservoir of Well JY1 was 160 ± 13 Ma, and that of the Wufeng Formation was 133 ± 15 Ma. The results of this experiment also have good correspondence. The time for the calcite vein to capture fluid is the same as that of the oil-gas secondary pyrolysis gas stage. The formation time of calcite veins is the same as that of a large amount of secondary pyrolysis gas. The formation of the calcite vein is accompanied by the opening of fractures. It can be seen from the burial history of the reservoir that when the temperature was about 190 °C, the formation was still in the stage of continuous increase with burial depth, and the pressure of the overlying formation was continuously increasing (Figure 9), and the fracture opened at this time, indicating that the opening of the fracture was caused by hydrocarbon generation. This time is 120 Ma earlier than the activity time of the Yanshan movement [28], so it is considered that overpressure is the key factor for fracture opening (Figure 9).

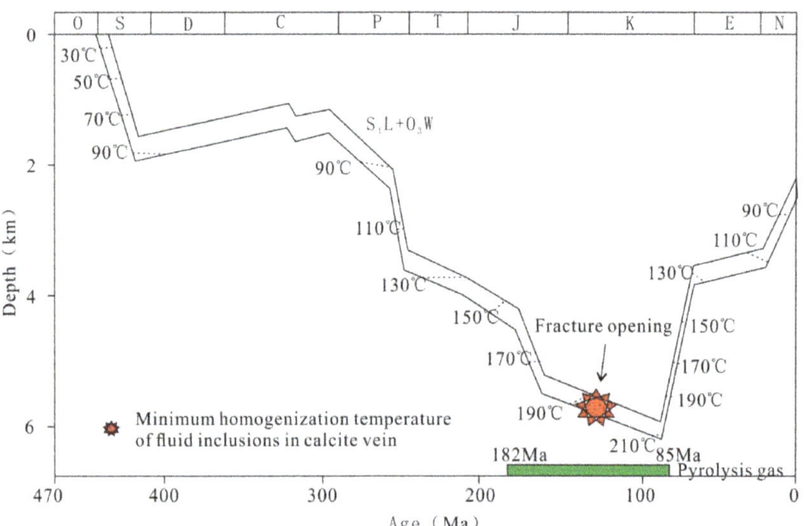

Figure 9. Comparison of minimum homogenization temperature and burial history of calcite vein fluid inclusions in Well JY1 (Modified from Gao et al. [12]).

5. Conclusions

The sedimentary environment controls the difference in the content of organic matter and brittle minerals in the shale reservoir. The parts with high content of organic matter and brittle minerals in the shale reservoir have a high development degree of microfracture.

According to the principle of material mechanics, the Silurian Longmaxi Formation and Ordovician Wufeng Formation are both located in the middle and lower part of the fold mechanics. Under compressive stress, which are powerful external causes for the formation of shale bedding joints, shale bedding is easy to form low-angle slip joints, interlayer slip joints, and interlayer foliation joints.

The time for the calcite vein to the formation and capture fluid is the same as that of the oil-gas secondary pyrolysis gas stage, which was accompanied by the opening of fractures. The stress analysis of the settlement process infers that the opening of the fracture was caused by hydrocarbon generation.

The formation of bedding fractures is jointly controlled by sedimentation, hydrocarbon generation, and tectonic movement. Sedimentation has formed shale with a complex bedding structure. The tectonic movement has created a stressful environment and space for the bedding fractures and promoted the opening of the bedding fractures. Excess silicon, organic acids, and the oil-gas produced in the hydrocarbon generation process promote and consolidate the bedding fractures.

Author Contributions: Conceptualization, Z.H. and S.J.; Methodology, Y.Z.; Project administration, Z.H.; Resources, H.N.; Software, Y.L.; Visualization, H.B.; Writing—review & editing, H.W. All authors have read and agreed to the published version of the manuscript.

Funding: This research is sponsored by National Natural Science Foundation of China (No. 41202103, No. 41902127) and China Postdoctoral Science Foundation (No. 2021M703000): Formation mechanism of compressive fracture network constrained by weak surface of shale bedding.Thanks would go to Editor Yong Li, and three anonymous reviewers for their constructive comments and suggestions.

Acknowledgments: We thank the Sinopec Exploration Company and the Sinopec Jianghan Oilfield, for the valuable data and information. We also thank the Sinopec management for permission to publish this work.

Conflicts of Interest: The authors declare no conflict of interest.

References

1. Zou, C.N.; Qiu, Z. New advances in unconventional petroleum sedimentology in China. *Acta Sedimentol. Sin.* **2021**, *39*, 1–9. (In Chinese with English Abstract)
2. Ou, C.H.; Li, C.C. 3D discrete network modeling of shale bedding fractures based on lithofacies characterization. *Pet. Explor. Dev.* **2017**, *44*, 336–345. (In Chinese with English Abstract) [CrossRef]
3. Shin, D.; Sharma, M. Factors Controlling the Simultaneous Propagation of Multiple Competing Fractures in a Horizontal Well. In Proceedings of the SPE Hydraulic Fracturing Technology Conference, The Woodlands, TX, USA, 4–6 February 2014. Paper SPE-168599.
4. Sun, J.; Bao, H.Y. Comprehensive characterization of shale gas reservoirs: A case study from Fuling shale gas field. *Pet. Geol. Exp.* **2018**, *40*, 1–12. (In Chinese with English Abstract)
5. Liang, C.; Jiang, Z.X.; Yang, Y.T.; Wei, X. Characteristics of shale lithofacies and reservoir space of the Wufeng–Longmaxi Formation, Sichuan Basin. *Pet. Explor. Dev.* **2012**, *39*, 691–698. [CrossRef]
6. Ding, W.L.; Zeng, W.T.; Wang, R.Y.; Wang, Z.; Sun, Y.; Wang, X. Method and application of tectonic stress field simulation and fracture distribution prediction in shale reservoir. *Earth Sci. Front.* **2016**, *23*, 63–74. (In Chinese with English Abstract) [CrossRef]
7. Zeng, L.B.; Li, Y.; Zhang, W.Y.; Liu, Y.Z.; Dong, G.P.; Shao, Q. The Effect of Multi-Scale Faults and Fractures on Oil Enrichment and Production in Tight Sandstone Reservoirs: A Case Study in the Southwestern Ordos Basin, China. *Front. Earth Sci.* **2021**, *9*, 664629. [CrossRef]
8. He, Z.L.; Hu, Z.Q.; Nie, H.K.; Li, S.; Xu, J. Characterization of shale gas enrichment in the Wufeng-Longmaxi Formation in the Sichuan Basin and its evaluation of geological construction-transformation evolution sequence. *Nat. Gas Geosci.* **2017**, *28*, 724–733. (In Chinese with English Abstract) [CrossRef]
9. Zhou, T.; Wang, H.B.; Li, F.X.; Li, Y.; Zou, Y.; Zhang, C. Numerical simulation of hydraulic fracture propagation in laminated shale reservoirs. *Pet. Explor. Dev.* **2020**, *47*, 1039–1051. (In Chinese with English Abstract) [CrossRef]
10. Cheng, Y.M. Impacts of the number of perforation clusters and cluster spacing on shale-gas wells. *SPE Reserv. Eval. Eng.* **2012**, *15*, 31–40. [CrossRef]
11. Li, F.X.; Huang, Z.W.; Ji, G.F.; Chen, S.; Zhou, T. Experimental study on the influence of shale fracture network structure and fluid on conductivity. *Unconv. Oil Gas* **2021**, *8*, 40–45. (In Chinese with English Abstract) [CrossRef]
12. Gao, J.; Zhang, J.; He, S.; Zhao, J.; He, Z.; Wo, Y.; Feng, Y.; Li, W. Overpressure generation and evolution in Lower Paleozoic gas shales of the Jiaoshiba region, China: Implications for shale gas accumulation. *Mar. Pet. Geol.* **2019**, *102*, 844–859. [CrossRef]
13. Murray, R.W.; Jones, D.L.; Brink, M. Diagenetic formation of bedded chert: Evidence from chemistry of the chert-shale couplet. *Geology* **1992**, *20*, 271–274. [CrossRef]
14. Xi, Z.; Tang, S.; Lash, G.G.; Ye, Y.; Lin, D.; Zhang, B. Depositional controlling factors on pore distribution and structure in the lower Silurian Longmaxi shales: Insight from geochemistry and petrology. *Mar. Pet. Geol.* **2021**, *130*, 105114. [CrossRef]
15. Li, Y.; Pan, S.; Ning, S.; Shao, L.; Jing, Z.; Wang, Z. Coal measure metallogeny: Metallogenic system and implication for resource and environment. *Sci. China Earth Sci.* **2022**, *65*, 1211–1228. [CrossRef]
16. Zeeb, C.; Gomez Rivas, E.; Bons, P.D.; Blum, P. Evaluation of Sampling Methods for Fracture Network Characterization Using Outcrops. *AAPG Bull.* **2013**, *97*, 1545–1566. [CrossRef]
17. Zeng, L.; Lyu, W.; Li, J.; Zhu, L.; Weng, J.; Jue, F.; Zu, K. Natural fractures and their influence on shale gas enrichment in Sichuan Basin, China. *J. Nat. Gas Sci. Eng.* **2016**, *30*, 1–9. [CrossRef]
18. Li, Y.; Chen, J.; Elsworth, D.; Pan, Z.; Ma, X. Nanoscale mechanical property variations concerning mineral composition and contact of marine shale. *Geosci. Front.* **2022**, *13*, 101405. [CrossRef]
19. Yue, F.; Jiao, W.; Guo, S. Controlling factors of fracture distribution of shale in Lower Cambrian Niutitang Formation in southeast Chongqing. *Coal Geol. Explor.* **2015**, *43*, 39–44.
20. Jin, Z.J.; Wang, G.P.; Liu, G.X.; Gao, B.; Liu, Q.; Wang, H.; Liang, X.; Wang, R. Research progress and keys scientific issues of continental shale oil in China. *Acta Pet. Sin.* **2021**, *42*, 821–835. (In Chinese with English Abstract)
21. Zhang, Z.Y.; Wei, Y.J.; Wei, W. Study on the types of coalbed gas deposits based on the fold neutral plane. *J. Saf. Environ.* **2015**, *15*, 153–157.
22. Zuo, J.P.; Li, Y.L.; Liu, C.; Liu, H.Y.; Wang, J.; Li, H.T.; Liu, L. Meso-fracture mechanism and its fracture toughness analysis of Longmaxi shale including different angles by means of M-SENB tests. *Eng. Fract. Mech.* **2019**, *215*, 178–192. [CrossRef]
23. Li, Y.; Chen, J.; Yang, J.; Liu, J.; Tong, W. Determination of shale macroscale modulus based on microscale measurement: A case study concerning multiscale mechanical characteristics. *Pet. Sci.* **2022**, *19*, 1262–1275. [CrossRef]
24. Zhao, J.; Jin, Z.; Jin, Z.; Geng, Y.; Wen, X.; Yan, C. Applying sedimentary geochemical proxies for paleoenvironment interpretation of organic-rich shale deposition in the Sichuan Basin, China. *Int. J. Coal Geol.* **2016**, *163*, 52–71. [CrossRef]
25. Zhu, H.; Ju, Y.; Yang, M.; Huang, C.; Feng, H.; Qiao, P.; Ma, C.; Su, X.; Lu, Y.; Shi, E.; et al. Grain-scale petrographic evidence for distinguishing detrital and authigenic quartz in shale: How much of a role do they play for reservoir property and mechanical characteristic? *Energy* **2022**, *239*, 122176. [CrossRef]
26. Wang, H.; He, Z.; Zhang, Y.; Bao, H.; Su, K.; Shu, Z.; Zhao, C.; Wang, R.; Wang, T. Dissolution of marine shales and its influence on reservoir properties in the Jiaoshiba area, Sichuan Basin, China. *Mar. Pet. Geol.* **2019**, *102*, 292–304. [CrossRef]

27. Gao, J.; He, S.; Yi, J. Discovery of high density methane inclusions in Jiaoshiba shale gas field and its significance. *Oil Gas Geol.* **2015**, *36*, 472–480. (In Chinese with English Abstract)
28. Liu, P.; Zhang, T.; Xu, Q.; Wang, X.; Liu, C.; Guo, R.; Lin, H.; Yan, M.; Qin, L.; Li, Y. Organic matter inputs and depositional palaeoenvironment recorded by biomarkers of marine-terrestrial transitional shale in the Southern North China Basin. *Geol. J.* **2022**, *57*, 1617–1627. [CrossRef]

Article

Differentiation and Prediction of Shale Gas Production in Horizontal Wells: A Case Study of the Weiyuan Shale Gas Field, China

Lixia Kang [1,2,*], Wei Guo [1,2,*], Xiaowei Zhang [1,2,*], Yuyang Liu [1,2] and Zhaoyuan Shao [1,2]

[1] PetroChina Research Institute of Petroleum Exploration and Development, Beijing 100083, China
[2] China National Shale Gas Research and Development (Experiment) Center, Langfang 065007, China
* Correspondence: kanglixia@petrochina.com.cn (L.K.); guowei69@petrochina.com.cn (W.G.); zhangxw69@petrochina.com.cn (X.Z.)

Abstract: The estimated ultimate recovery (EUR) of shale gas is an important index for evaluating the production capacity of horizontal wells. The Weiyuan shale gas field has wells with considerable EUR differentiation, which hinders the prediction of the production capacity of new wells. Accordingly, 121 wells with highly differentiated production are used for analysis. First, the main control factors of well production are identified via single-factor and multi-factor analyses, with the EUR set as the production capacity index. Subsequently, the key factors are selected to perform the multiple linear regression of EUR, accompanied by the developed method for well production prediction. The thickness and drilled length of Long $1_1{}^1$ (Substratum 1 of Long 1 submember, Lower Silurian Longmaxi Formation) are demonstrated to have the uttermost effects on the well production, while several other factors also play important roles, including the fractured horizontal wellbore length, gas saturation, brittle mineral content, fracturing stage quantity, and proppant injection intensity. The multiple linear regression method can help accurately predict EUR, with errors of no more than 10%, in wells that have smooth production curves and are free of artificial interference, such as casing deformation, frac hit, and sudden change in production schemes. The results of this study are expected to provide certain guiding significances for shale gas development.

Keywords: shale gas; grey correlation method; multiple linear regression; production evaluation; main control factor; estimated ultimate recovery

Citation: Kang, L.; Guo, W.; Zhang, X.; Liu, Y.; Shao, Z. Differentiation and Prediction of Shale Gas Production in Horizontal Wells: A Case Study of the Weiyuan Shale Gas Field, China. *Energies* **2022**, *15*, 6161. https://doi.org/10.3390/en15176161

Academic Editors: Yong Li, Fan Cui and Chao Xu

Received: 6 July 2022
Accepted: 1 August 2022
Published: 25 August 2022

Publisher's Note: MDPI stays neutral with regard to jurisdictional claims in published maps and institutional affiliations.

1. Introduction

The recovery technology for shallow (<3500 m) marine shale gas in China has been mature in terms of supporting techniques and fit-for-purpose equipment [1]. Starting from scratch, the shale gas industry of China reached an annual production of 1.0×10^{10} m^3 within six years, which was doubled to approach a historical breakthrough in eight years [2]. Shale gas reservoirs are found in various sedimentary environments in China [3], including marine (50% of the whole country's shale gas resources), marine–continental transitional (20%), and continental (30%) environments [4–7].

Located in the southern Sichuan Basin [8,9], the Weiyuan shale gas field has been in exploration and development since 2009, with a cumulative proven shale gas resource in place of 4.277×10^{11} m^3 [10]. In 2021, the shale gas production climbs up to 4.12×10^9 m^3, which accounts for 32% of the total shale gas production of China National Petroleum Corporation (CNPC). In this context, it is of critical importance to calculate the estimated ultimate recovery (EUR) of gas wells for accurately estimating the potential of shale gas recovery and realizing large-scale economic development.

EUR is an important index for evaluating the production capacity of a shale gas well, and it is affected by numerous factors [11,12], which can be analyzed by various methods [13,14]. Most popular methods include the grey correlation approach [15,16]

and numerical simulations [17]. Efforts have been made to quantitatively investigate the correlation between the engineering parameters and the fracture network effectiveness via comparison of reservoir attributes and analysis of well performances [18,19]. Moreover, numerous methods have been developed for calculating the EUR of shale gas wells, including the empirical method, the modern production decline analysis, the analytical method, the linear flow analysis, and the big data approach [20]. Specifically, the pressure transient and Blasingame production decline analysis were used to calculate the EUR of a well with short-term production [21]. Moreover, innovative models for the multi-stage fractured horizontal well [22] and those based on the deep neural network [23] are useful ways of predicting the production of shale gas wells. Niu et al. [24] forecast the production capacity of shale gas wells of the Weiyuan gas field via multiple regression. Furthermore, Yin et al. [25] developed a production capacity assessment model for shale gas wells by building and solving the dual-porosity mathematical model of shale gas reservoirs. Accurately calculating EUR of wells and analyzing the main control factors of the production capacity of shale gas wells are of great significance for improving well production and guiding the deployment of shale gas recovery.

In the Weiyuan shale gas field, the production of horizontal wells is highly differentiated from well to well, and thus systematic efforts should be made to clarify the main control factors of gas well production capacity. In this research, single-factor analysis and grey correlation analysis are used to identify key factors affecting the production capacity of wells, while multiple linear regression is used to predict the production capacity of a single well. The findings of this research are expected to guide shale gas well operation in the Weiyuan gas field and enable high-efficiency recovery of shale gas.

2. Geological Setting

The Weiyuan gas field is located in the Weiyuan and Zizhong Counties in the southwestern Sichuan Basin. It lies in the Southwest Sichuan gentle fold zone of the Central Sichuan uplift and presents itself as a large dome anticline (Figure 1). The main layers for shale gas recovery are within the Wufeng Formation–Longmaxi Formation. The Wufeng–Longmaxi shale is buried at 2000–4000 m, with a thickness of 180–600 m. The interval from the Wufeng Formation to the Long 1 Sub-Member of the Longmaxi Formation is the target interval, with a shale thickness of 43.9–54.8 m, averaging 46.4 m. The Long 1 Sub-Member is further divided into the Long $1_1{}^1$–$1_1{}^4$ small layers from the bottom to top, among which the Long $1_1{}^1$ is the main layer for recovery, with a thickness of 1.7–7.0 m (4.6 m on average). The shale in the interval from the Wufeng Formation to the Long 1 Sub-Member has a TOC of 2.7–3.6%, averaging 3.2%; a porosity of 5.2–6.7%, averaging 5.9%; a gas saturation of 32.7–84.6%, averaging 64.7%; a gas content of 3.3–8.5 m^3/t, averaging 5.5 m^3/t; a brittle mineral content of 60–82%, averaging 74%; and a pressure coefficient of 1.2–2.0 in the favorable zone for production capacity building.

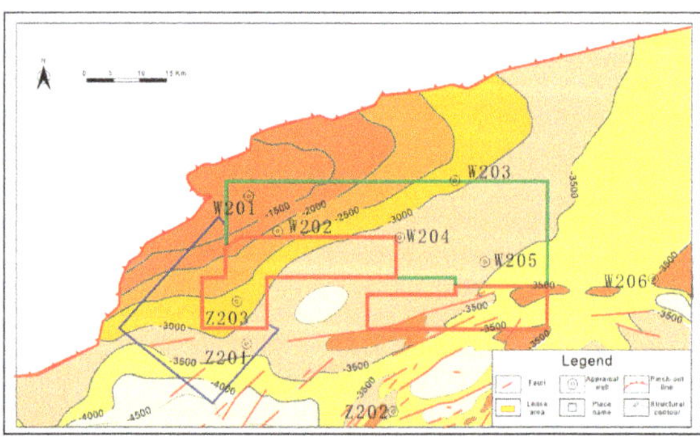

Figure 1. Structural map of the Weiyuan shale gas field.

3. Data and Analysis Methods

By the end of 2021, a total of 396 wells have been brought into production in the Weiyuan block and these wells present an average testing daily production of 2.16×10^5 m^3/d and an average single-well EUR of 8.9×10^7 m^3. A total of 121 wells that were brought into production within the recent three years and featured considerably differentiated production are selected as the samples for analysis, which present the average testing daily production of 2.488×10^5 m^3/d, the average first-year daily production of 9.4×10^4 m^3/d, and the EUR of $(0.31-2.22) \times 10^8$ m^3 (averaging 1.06×10^8 m^3; Figure 2; Table 1). Among these wells, low-production wells with the EUR lower than 0.6×10^8 m^3 account for 8%; medium-production wells with the EUR of $(0.6-1.2) \times 10^8$ m^3, 59%; and high-production wells with the EUR above 1.2×10^8 m^3, 33%. A total of 62 wells are found with EUR below 1.0×10^8 m^3, which account for about 50% of the wells brought into production. The overall distribution of EUR is even and the production of wells is greatly differentiated.

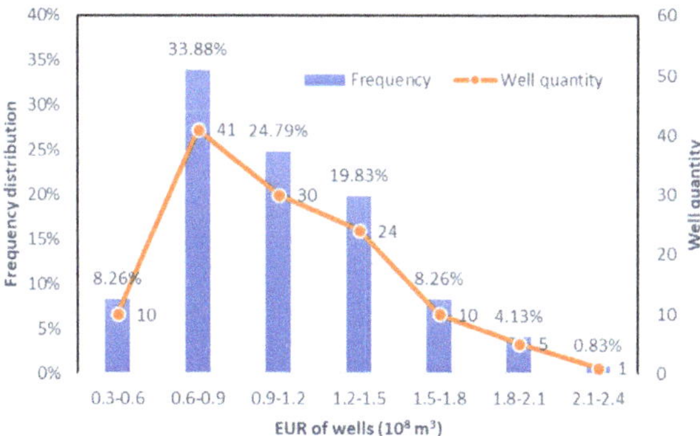

Figure 2. Probability distribution of the EUR of the 121 sample wells.

Table 1. Parameter statistics of the sample wells of the Weiyuan shale gas field.

Parameter	Unit	Sampling Quantity	Min	Max	Average	Standard Deviation
EUR	10^8 m^3	121	0.31	2.22	1.06	0.38
Long $1_1{}^1$ interval thickness	m	121	2.40	7.60	5.37	1.37
TOC	%	121	3.30	6.75	5.22	0.53
Brittle mineral content	%	121	63	96	83	5.43
Gas content	m^3/t	121	5.43	10.30	7.34	1.03
Gas saturation	%	121	71	84	76	3.60
Pressure coefficient	dimensionless	121	1.40	2.05	1.74	0.18
Drilled length of the Long $1_1{}^1$ interval	m	121	482	2515	1593	344.75
Fractured horizontal wellbore length	m	121	899	2577	1664	328.16
Fracturing stage quantity	Stages	121	16	36	23	4.36
Liquid injection intensity	m^3/m	121	16	48	27	4.13
Proppant injection intensity	t/m	121	1.08	3.50	2.08	0.52

3.1. Factors Affecting Gas Well Production

Various factors affect the production capacity of shale gas wells, which include geological and engineering factors. The geological factors include the thickness of the Long $1_1{}^1$ interval, TOC, brittle mineral content, gas content, gas saturation, and pressure coefficient, while the engineering factors include the drilled length of the Long $1_1{}^1$ interval, the fractured horizontal wellbore length, fracturing stage quantity, liquid injection intensity, and proppant injection intensity.

3.2. Grey Correlation Analysis

The production capacity of shale gas wells is subjected to the joint effects of multiple factors, which may interact with each other. The basic regression method fails to deliver a satisfactory analysis of the main control factors, as it cannot quantitatively capture the root causes of the differentiated production capacity of wells [26]. Hence, the grey correlation analysis is performed to quantify the intensity of impacts of each factor on the well production capacity and correspondingly identify the main control factors of the production capacity of shale gas horizontal wells in the Weiyuan gas field.

The grey system theory was developed by the Chinese scholar Julong Deng in 1982. This theory deals with the "limited-sample" system with poor data (lean in information) with partially known and partially unknown information [27]. The grey correlation analysis is an important analytical method for the grey system theory and is highly valuable in investigating the correlations among variables [28]. When no strict mathematical relationships exist between each influential factor and the total result, the grey correlation method can deliver an effective analysis of data and describe the strength, magnitude, and ranking of relationships between factors [29]. The grey correlation analysis determines the closeness of the relationship, according to the geometric similarity between the reference sequence and analysis sequence, and can be used to determine how intensive the effects of each factor are on the result [30].

The grey correlation analysis is performed after normalizing the above data of the factors. The temporal sequence of the EUR is set as the reference sequence X0, while the temporal sequences of each factor, namely the Long $1_1{}^1$ thickness, TOC, brittle mineral content, gas content, gas saturation, pressure coefficient, drilled Long $1_1{}^1$ length, fractured horizontal wellbore length, fracturing stage quantity, liquid injection intensity, and proppant injection intensity, are used as the analysis sequences. The following matrix is built:

$$(X0, X1, \cdots, Xn) = \begin{bmatrix} x_0(1) & x_1(1) & \cdots & x_n(1) \\ x_0(2) & x_1(2) & \cdots & x_n(2) \\ \vdots & \vdots & \ddots & \vdots \\ x_0(m) & x_1(m) & \cdots & x_n(m) \end{bmatrix} \tag{1}$$

Then, the absolute difference between the elements of each analysis sequence and the reference sequence is calculated one after another:

$$\Delta_{0i}(k) = |x_0(k) - x_i(k)|(k = 1, \cdots, m, i = 1, \cdots, n)$$

Accordingly, the absolute difference matrix is formed:

$$\begin{bmatrix} \Delta_{01}(1) & \Delta_{02}(1) & \cdots & \Delta_{0n}(1) \\ \Delta_{01}(2) & \Delta_{02}(2) & \cdots & \Delta_{0n}(2) \\ \vdots & \vdots & \ddots & \vdots \\ \Delta_{01}(m) & \Delta_{02}(m) & \cdots & \Delta_{0n}(m) \end{bmatrix} \tag{2}$$

The maximum value in the absolute difference matrix is defined as the maximum difference, while the minimum one is identified as the minimum difference:

$$\Delta(max) = \max\{\Delta_{0i}(k)\} \tag{3}$$

$$\Delta(min) = \min\{\Delta_{0i}(k)\} \tag{4}$$

The correlation coefficient is computed as below:

$$\xi_{0i}(k) = \frac{\Delta(min) + \rho\Delta(max)}{\Delta_{0i}(k) + \rho\Delta(max)} \tag{5}$$

where ρ is the resolution ratio determining the effects of $\Delta(max)$ on the data conversion; $\xi_{0i}(k)$ is the correlation coefficient with a positive value of no more than 1. The function of ρ is to control the influence of the $\Delta(max)$ on the overall correlation degree, so that the resolution of the correlation degree between different factors at the longitudinal level is as large as possible. A lower ρ results in promoted differentiation of correlation coefficients. In this research, ρ is empirically set as 0.5.

The grey correlation degrees between factors are computed:

$$\gamma_{0i} = \frac{1}{N} \sum_{k=1}^{N} \xi_{0i}(k) \ (i = 1,2,3,\cdots,n) \tag{6}$$

where γ_{0i} is the correlation degree, a measure of the correlation between the reference sequence X_0 and the analysis sequence $X_i (i = 1,2,3,\cdots,n)$.

The weight coefficients of each factor are determined using the following equation, according to the correlation degrees:

$$W_i = \frac{\gamma_{0i}}{\sum_{i=1}^{n} \gamma_{0i}} \tag{7}$$

3.3. Principles of the Multiple Regression Algorithm

The multiple linear regression (MLR) analysis is a mathematical analysis method based on the correlation between the dependent and independent variables [31,32]. With a dependent variable related to multiple factors, MLR can be performed for analysis [33], in which an MLR model is developed via the weighted summation of these factors [34]. It is assumed that the dependent variable y can be expressed as a linear combination of m independent variables:

$$y = \beta_0 + \beta_1 x_1 + \beta_2 x_2 + \cdots + \beta_i x_m + \varepsilon \tag{8}$$

where y is the dependent variable; x_1, $x_2 \cdots x_i$ are the independent variables; β_0, β_1, $\beta_2 \cdots \beta_m$ are the regression coefficients; and ε is the random error with the normal distribution $N(0, \delta^2)$.

Providing that there are N sets of observation data (samples), the theoretical equation shall be:

$$\mathbf{y} = \mathbf{X}\beta + \epsilon \tag{9}$$

where:

$$\mathbf{y} = \begin{bmatrix} y_1 \\ y_2 \\ \vdots \\ y_n \end{bmatrix} \tag{10}$$

$$\mathbf{X} = \begin{bmatrix} 1 & x_{11} & \cdots & x_{1m} \\ 1 & x_{21} & \cdots & x_{2m} \\ \vdots & \vdots & \vdots & \vdots \\ 1 & x_{n1} & \cdots & x_{nm} \end{bmatrix} \tag{11}$$

$$\epsilon = \begin{bmatrix} \epsilon_1 \\ \epsilon_2 \\ \vdots \\ \epsilon_n \end{bmatrix} \tag{12}$$

$$\beta = \begin{bmatrix} \beta_0 \\ \beta_1 \\ \vdots \\ \beta_m \end{bmatrix} \tag{13}$$

In accordance with the least-square principle, the difference between the regression estimate and the true value (also known as the residual error) is calculated and the regression coefficients are computed by searching the least sum of squares of residual errors. Hence, the regression equation is obtained.

The test of significance can be performed for the regression coefficients of the MLR equation to validate whether or not a linear relationship exists between variables. On one hand, the regression coefficients that are not all zero stand for significant linear relationships between variables. On the other hand, regression coefficients that are all zero represent no significant linear relationship.

4. Results and Discussion

4.1. Single-Factor Analysis

As is shown in Figure 3, EUR is found with relatively large positive correlations with the Long $1_1{}^1$ thickness, gas saturation, drilled length of the Long $1_1{}^1$, fractured horizontal wellbore length, and fracturing stage quantity, whereas it presents smaller correlations with TOC, brittle mineral content, gas content, pressure coefficient, liquid injection intensity, and proppant injection intensity.

The Pearson correlation analysis is implemented on EUR and 11 influential factors after the Z-score normalization to reveal the linear relationships among these factors and those between each factor and the EUR (Table 2 and Figure 4). Significant correlations are seen between EUR and Long $1_1{}^1$ thickness, brittle mineral content, gas saturation, drilled Long $1_1{}^1$ length, fractured horizontal wellbore length, fracturing stage quantity, and proppant injection intensity, whereas weak relationships are observed between EUR and TOC, gas content, pressure coefficient, and liquid injection intensity.

Table 2. Results of the Pearson correlation analysis.

		EUR	Long $1_1{}^1$ Interval Thickness	TOC	Brittle Mineral Content	Gas Content	Gas Saturation	Pressure Coefficient	Drilled Length of the Long $1_1{}^1$ Interval	Fractured Horizontal Wellbore Length	Fracturing Stage Quantity	Liquid Injection Intensity	Proppant Injection Intensity
EUR	Correlation	1.00	0.663 **	(0.08)	0.325 **	(0.14)	0.482 **	0.14	0.526 **	0.499 **	0.298 **	0.08	0.292 **
	Significance		0.00	0.37	0.00	0.12	0.00	0.14	0.00	0.00	0.00	0.37	0.00
Long $1_1{}^1$ Interval Thickness	Correlation	0.663 **	1.00	0.10	0.338 **	0.03	0.709 **	0.262 **	0.524 **	0.494 **	0.08	−0.188 *	0.481 **
	Significance	0.00		0.30	0.00	0.73	0.00	0.00	0.00	0.00	0.39	0.04	0.00
TOC	Correlation	(0.08)	0.10	1.00	(0.15)	0.584 **	0.286 **	0.435 **	(0.17)	(0.16)	(0.10)	0.10	0.246 **
	Significance	0.37	0.30		0.10	0.00	0.00	0.00	0.06	0.09	0.27	0.29	0.01
Brittle Mineral Content	Correlation	0.325 **	0.338 **	(0.15)	1.00	−0.190 *	0.250 **	0.224 *	0.282 **	0.12	0.09	(0.06)	(0.06)
	Significance	0.00	0.00	0.10		0.04	0.01	0.01	0.00	0.18	0.31	0.54	0.51
Gas Content	Correlation	(0.14)	0.03	0.584 **	−0.190*	1.00	0.03	0.595 **	(0.12)	(0.06)	(0.12)	(0.04)	0.232*
	Significance	0.12	0.73	0.00	0.04		0.77	0.00	0.19	0.53	0.19	0.64	0.01
Gas Saturation	Correlation	0.482 **	0.709 **	0.286 **	0.250 **	0.03	1.00	0.18	0.316 **	0.363 **	0.11	(0.09)	0.421 **
	Significance	0.00	0.00	0.00	0.01	0.77		0.05	0.00	0.00	0.23	0.33	0.00
Pressure Coefficient	Correlation	0.14	0.262 **	0.435 **	0.224 *	0.595 **	0.18	1.00	0.01	(0.11)	(0.04)	(0.04)	(0.13)
	Significance	0.14	0.00	0.00	0.01	0.00	0.05		0.90	0.21	0.69	0.68	0.17
Drilled Length of the Long $1_1{}^1$ Interval	Correlation	0.526 **	0.524 **	(0.17)	0.282 **	(0.12)	0.316 **	0.01	1.00	0.753 **	0.348 **	(0.17)	0.336 **
	Significance	0.00	0.00	0.06	0.00	0.19	0.00	0.90		0.00	0.00	0.06	0.00
Fractured Horizontal Wellbore Length	Correlation	0.499 **	0.494 **	(0.16)	0.12	(0.06)	0.363 **	(0.11)	0.753 **	1.00	0.534 **	−0.24 **	0.403 **
	Significance	0.00	0.00	0.09	0.18	0.53	0.00	0.21	0.00		0.00	0.01	0.00
Fracturing Stage Quantity	Correlation	0.298 **	0.08	(0.10)	0.09	(0.12)	0.11	(0.04)	0.348 **	0.534 **	1.00	0.35 **	(0.17)
	Significance	0.00	0.39	0.27	0.31	0.19	0.23	0.69	0.00	0.00		0.00	0.06
Liquid Injection Intensity	Correlation	0.08	−0.188 *	0.10	(0.06)	(0.04)	(0.09)	(0.04)	(0.17)	−0.24 **	0.35 **	1.00	(0.11)
	Significance	0.37	0.04	0.29	0.54	0.64	0.33	0.68	0.06	0.01	0.00		0.22
Proppant Injection Intensity	Correlation	0.292 **	0.481 **	0.246 **	(0.06)	0.232 *	0.421 **	(0.13)	0.336 **	0.403 **	(0.17)	(0.11)	1.00
	Significance	0.00	0.00	0.01	0.51	0.01	0.00	0.17	0.00	0.00	0.06	0.22	

Notes: * represents a significant correlation; ** represents an extremely significant correlatio.

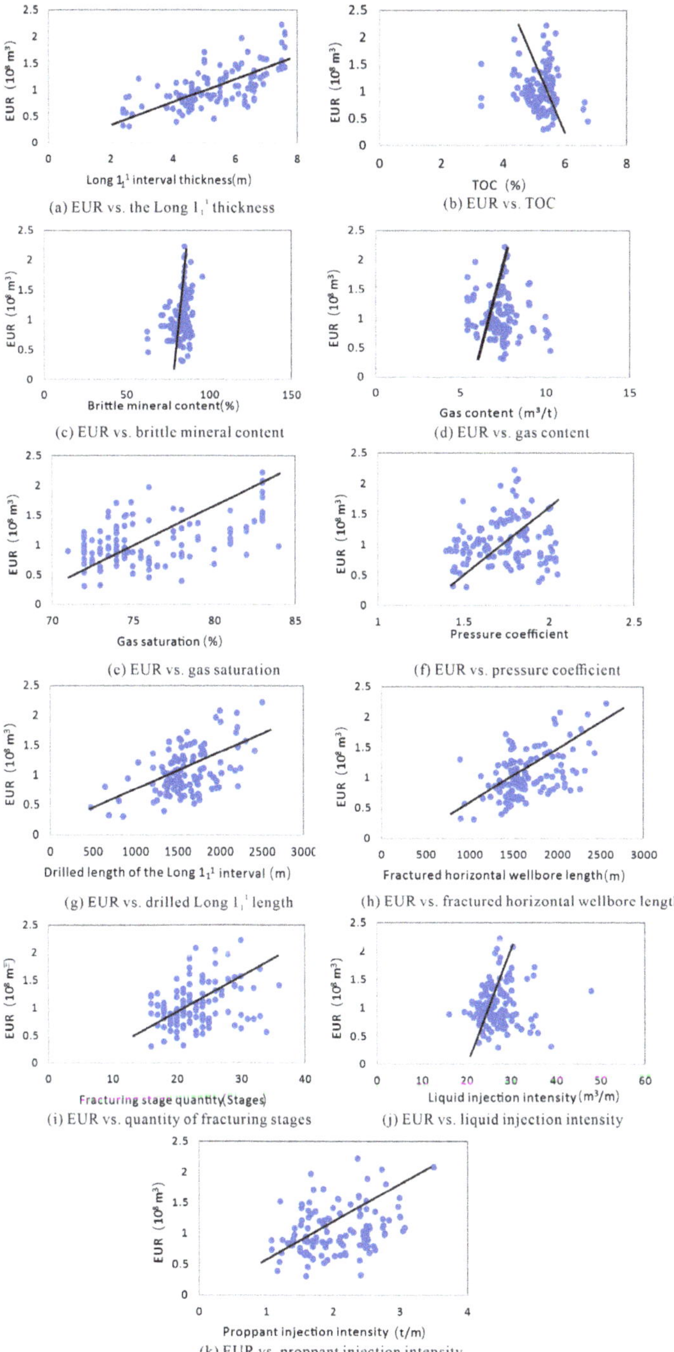

Figure 3. Single-factor regression of factors affecting EUR.

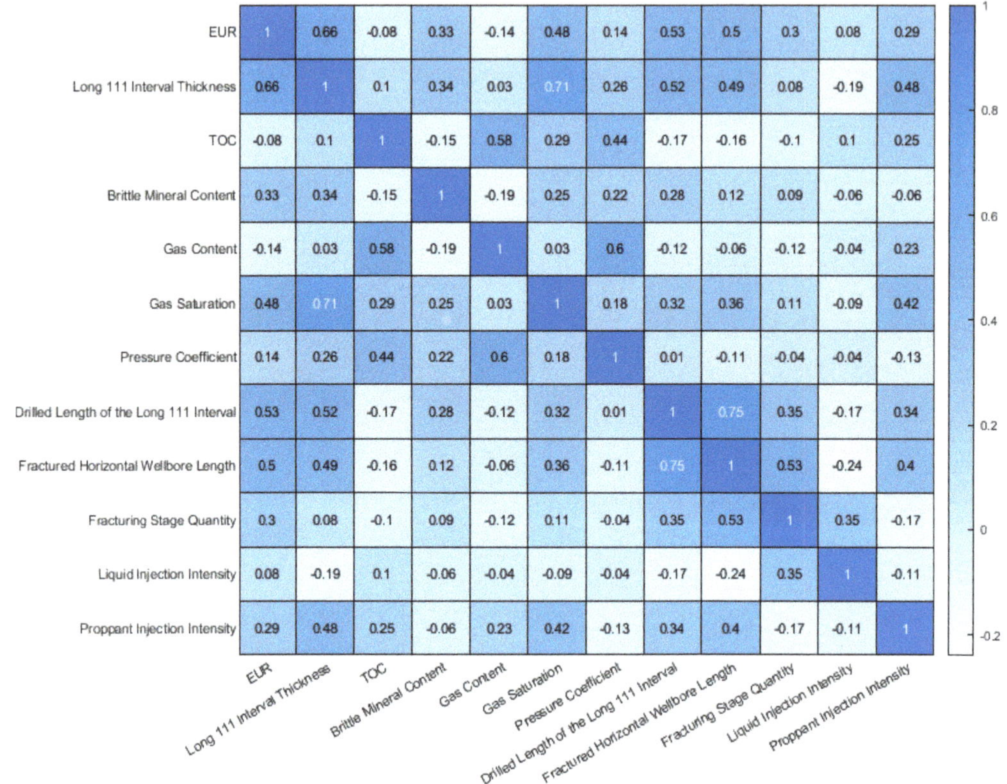

Figure 4. Heat map of the single-factor analysis results.

4.2. Grey Correlation Analysis

The correlation degrees between each factor and EUR are shown in Table 3, while the corresponding weight coefficients are presented in Figure 5. Clearly, the thickness and drilled length of Long $1_1{}^1$ have the highest effects on the EUR, followed by the second tier composed of the fractured horizontal wellbore length, gas saturation, brittle mineral content, and fracturing stage quantity, and the third tier consisting of the proppant injection intensity, liquid injection intensity, TOC, pressure coefficient, and gas content. According to the results, for the first seven factors, the results are consistent with single-factor analysis and grey correlation analysis, which indicates that the results are reliable.

Table 3. Calculated correlation degrees between each factor and the EUR.

Factor	Long $1_1{}^1$ Interval Thickness	TOC	Brittle Mineral Content	Gas Content	Gas Saturation	Pressure Coefficient	Drilled Length of the Long $1_1{}^1$ Interval	Fractured Horizontal Wellbore Length	Fracturing Stage Quantity	Liquid Injection Intensity	Proppant Injection Intensity
Correlation Degree	0.810	0.737	0.771	0.729	0.774	0.730	0.790	0.786	0.766	0.744	0.744
Ranking	1	9	5	11	4	10	2	3	6	8	7

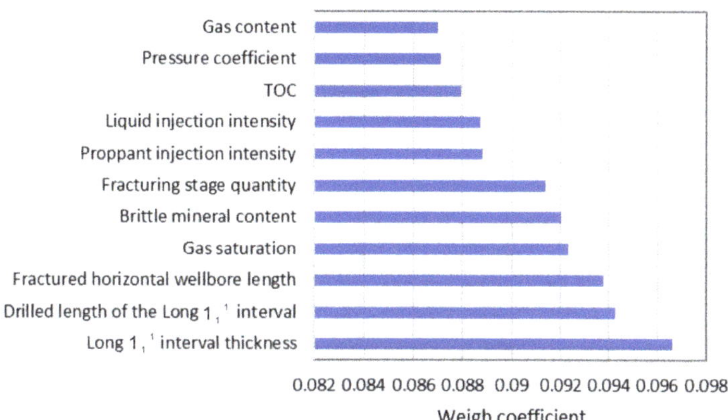

Figure 5. Weight coefficients characterizing the effects of each factor on the EUR of wells.

4.3. MLR Analysis

According to the results of the grey correlation analysis of the factors affecting well production, seven key factors that are highly correlated with the EUR are identified and selected for the MLR analysis. These factors are the Long 1_1^1 thickness, brittle mineral content, gas saturation, drilled Long 1_1^1 length, fractured horizontal wellbore length, fracturing stage quantity, and proppant injection intensity.

As shown in Table 4, the calculated significance is 1.1×10^{-15}, far lower than the 0.05 significance level, and the constant for the MLR is non-zero, which means that the regression performance of the MLR equation is satisfactory and the results of the linear regression analysis between the EUR and these factors are reliable. The MLR fitting formula of the EUR of a shale gas horizontal well in relation to the above factors can be written as(Table 5): the EUR of a shale gas well = $-1.008 + 0.143 \times$ Long 1_1^1 thickness + $0.006 \times$ brittle mineral content + $0.001 \times$ gas saturation + $1.6 \times 10^{-4} \times$ drilled Long 1_1^1 length $- 4.9 \times 10^{-6} \times$ fractured horizontal wellbore length + $0.018 \times$ fracturing stage quantity + $0.022 \times$ proppant injection intensity.

Table 4. Variance analysis of the MLR model.

Variance Analysis	Sum of Squares	Degree of Freedom	Mean Square	F	Significance
Regression	8.933	7	1.276	17.697	1.10×10^{-15}
Residual	8.148	113	0.072		
Total	17.081	120			

Table 5. Results of the MLR calculation.

Constant	Long 1_1^1 Interval Thickness	Brittle Mineral Content	Gas Saturation	Drilled Long 1_1^1 Length	Fractured Horizontal Wellbore Length	Fracturing Stage Quantity	Proppant Injection Intensity
-1.008	0.143	0.006	0.001	1.6×10^{-4}	-4.9×10^{-6}	0.018	0.022

4.4. Application

The EUR of 20 shale gas wells of the Weiyuan gas field is calculated using the MLR production capacity model presented above and compared with that computed using the analytical method. The analytical method considers the shale reservoir's physical properties, fracture parameters, fracture conductivity, and other model parameters. Moreover, it establishes the shale gas horizontal well model, with the model parameters historically fitted to correct the model parameters. After the historical fitting is completed, the production is predicted. Generally, the analytical method is used to calculate EUR for shale gas and the results are reliable in most cases. The results show (Table 6) that production predictions of 4 wells are overestimated, those of 5 are underestimated, and those of 11 are found with relatively accurate estimates, with errors no more than 10%. The matching rate of the model is around 50%.

Table 6. Comparison of the results between MLR and the analytical method.

Prediction Performance	Well No.	Testing Daily Production (10^4 m^3/d)	First-Year Daily Production (10^4 m^3/d)	EUR Based on the Analytical Method (10^8 m^3)	EUR Based on MLR (10^8 m^3)	Relative Error
Overestimation	W1	14.37	5.14	0.64	0.84	30.3%
	W2	24.93	11.54	0.98	1.25	27.5%
	W3	28.57	10.43	1.08	1.20	10.8%
	W4	28.56	10.42	0.90	1.00	10.5%
Accurate Estimation	W5	33.12	13.15	1.41	1.54	9.4%
	W6	13.98	6.42	0.89	0.96	8.4%
	W7	17.58	6.84	0.87	0.89	3.0%
	W8	43.55	14.54	1.58	1.60	1.4%
	W9	16.41	7.79	0.90	0.90	0.7%
	W10	31.43	13.88	1.36	1.36	0.0%
	W11	17.45	8.40	0.97	0.97	-0.3%
	W12	21	8.28	0.90	0.88	-2.2%
	W13	32.65	11.83	1.35	1.31	-3.1%
	W14	21.04	13.50	1.30	1.20	-8.0%
	W15	22.38	10.72	1.32	1.19	-9.7%
Underestimation	W16	28.19	10.32	1.08	0.88	-18.6%
	W17	31.01	13.02	1.31	1.05	-20.1%
	W18	35.55	19.42	1.71	1.28	-25.2%
	W19	31.03	10.72	1.52	1.10	-27.9%
	W20	40.45	16.80	1.97	1.35	-31.3%

Furthermore, the production performance of the wells presenting accurate estimates (Figure 6) are typically found with continuous production and pressure curves, with no artificial interference at the time being brought into production and during the production (e.g., engineering-induced casing deformation, frac hit, and sudden change in the production scheme). In contrast, the typical production performances of the overestimated wells (Figure 7) reveal that the wells are subjected to frac hit from the adjacent well and present discontinuous production curves; accordingly, the actual EUR far deviates from the ideal EUR predicted by the MLR model. Moreover, the typical production performances of the underestimated wells (Figure 8) show a change in the production scheme around the 350th day of production, and the production rate and pressure remain stable for a rather long period; as a result, the predicted EUR is lower than the actual EUR. The above analysis suggests that effective screening of sample data is required to select wells with continuous production decline for the MLR analysis in an attempt to avoid artificial interference and improve the accuracy of the MLR model.

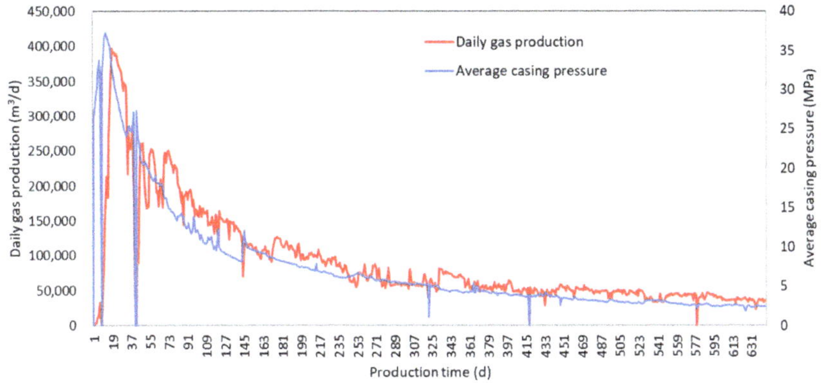

Figure 6. The production history of Well W5 (with the accurate prediction).

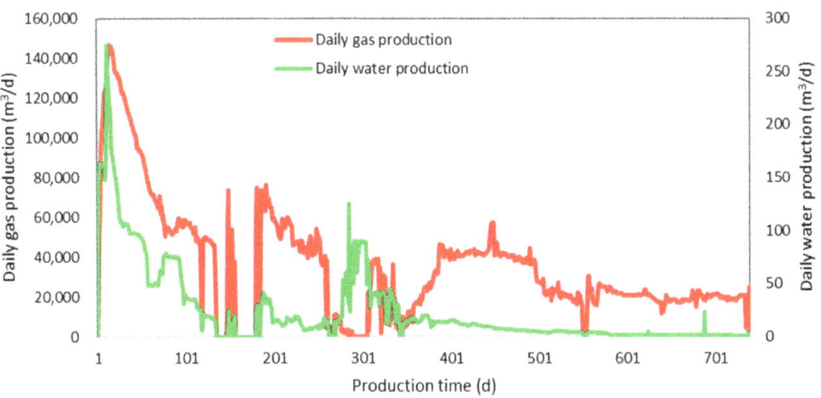

Figure 7. The production history of Well W1 (overestimated).

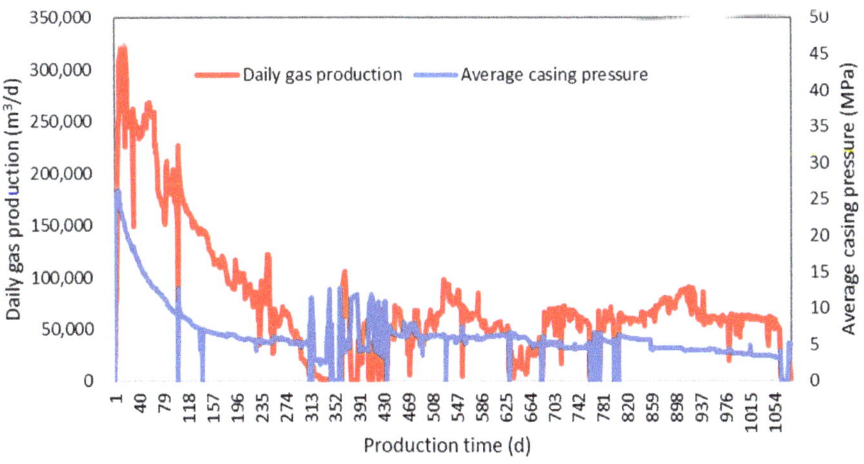

Figure 8. The production history of Well W17 (underestimated).

5. Conclusions

(1) First, the main geological and engineering factors that control the production capacity of horizontal wells in the Weiyuan shale gas field are clarified via the single-factor analysis and grey correlation analysis, with 121 horizontal wells used as the samples. The primary control factors are the thickness and drilled length of Long $1_1{}^1$, while the secondary control factors are the fractured horizontal wellbore length, gas saturation, brittle mineral content, fracturing stage quantity, and proppant injection intensity.

(2) For the production forecast of wells based on the multiple linear regression, one needs to select wells with continuous production curves and that are free of artificial interference when being brought into production or during production, such as casing deformation, frac hit, and sudden change in production schemes. For such wells, the model presented in this research can accurately predict the EUR of wells. For the wells newly brought into production, the model presented in this research can rapidly predict the ideal production capacity and the EUR of wells, with errors of no more than 10% compared to the analytical results.

Author Contributions: Conceptualization, L.K., W.G. and X.Z.; Data curation, L.K., W.G., Y.L. and Z.S.; Formal analysis, X.Z., Y.L. and Z.S.; Investigation, L.K., W.G., Y.L. and Z.S.; Methodology, L.K., X.Z. and Y.L.; Project administration, W.G. and X.Z. All authors have read and agreed to the published version of the manuscript.

Funding: This research was funded by [Study on productivity evaluation and development technology policy optimization of Marine and continental transitional shale gas] grant number [2021DJ2005] and [Study on shale gas optimization and scale efficiency development technology in southern Sichuan] grant number [2022KT1204].

Institutional Review Board Statement: Not applicable.

Informed Consent Statement: Not applicable.

Data Availability Statement: Not applicable.

Conflicts of Interest: The authors declare no conflict of interest.

References

1. Xu, C.H.; Zhong, Y.H.; Wang, C.S.; Ke, L.; Chou, W.D.; Zhao, Q. A brief analysis of key points and early production characteristics of shale gas exploration and development in China. *J. Geol.* **2021**, *45*, 197–212.
2. Zou, C.N.; Zhao, Q.; Cong, L.Z.; Wang, H.Y.; Shi, Z.S.; Wu, J.; Pan, S.Q. Development progress, potential and prospect of shale gas in China. *Nat. Gas Ind.* **2021**, *41*, 1–14.
3. Feng, Y.; Ning, Z.F.; Liu, H.Q. Fractal characteristics of shales from a shale gas reservoir in the Sichuan Basin, China. *Fuel* **2014**, *115*, 378–384.
4. Zou, C.N.; Yang, Z.; Sun, S.S.; Zhao, Q.; Bai, W.H.; Liu, H.L.; Pan, S.Q.; Wu, S.T.; Yuan, Y.L. Exploring petroleum inside source kitchen: Shale oil and gas in Sichuan Basin. *Sci. China Earth Sci.* **2020**, *50*, 903–920. [CrossRef]
5. Zhao, W.Z.; Jia, A.L.; Wei, Y.S.; Wang, J.L.; Zhu, H.Q. Progress in shale gas exploration in China and prospects for future development. China. *Pet. Explor.* **2020**, *25*, 31–44.
6. Li, Y.; Chen, J.-Q.; Yang, J.-H.; Liu, J.-S.; Tong, W.-S. Determination of shale macroscale modulus based on microscale measurement: A case study concerning multiscale mechanical characteristics. *Pet. Sci.* **2022**, *19*, 1262–1275. [CrossRef]
7. Li, Y.; Wang, Z.; Pan, Z.; Niu, X.; Meng, S. Pore structure and its fractal dimensions of transitional shale: A cross-section from east margin of the ordos basin, china. *Fuel* **2019**, *241*, 417–431. [CrossRef]
8. Liang, F.; Zhang, Q.; Lu, B.; Jiang, W.; Xiong, X.L.; Chen, P.; Jiang, R.; Ma, C. Lithofacies and distribution of Wufeng-Longmaxi organic-rich shale and its impact on shale gas production in Weiyuan shale gas play, southern Sichuan Basin, China. *Acta Sedimentol. Sin.* **2022**; in press.
9. Ma, X.H.; Li, X.Z.; Liang, F.; Wan, Y.J.; Shi, Q.; Wang, Y.H.; Zhang, X.W.; Che, M.G.; Guo, W.; Guo, W. Dominating factors on well productivity and development strategies optimization in Weiyuan shale gas play, Sichuan Basin, SW China. *Pet. Explor. Dev.* **2020**, *47*, 555–563. [CrossRef]
10. He, X.; Wu, J.F.; Yong, R.; Zhao, S.X.; Zhou, X.J.; Zhang, J.D.; Zhang, D.L.; Zhong, C.X. Accumulation conditions and key exploration and development technologies of marine shale gas field in Changning-Weiyuan block, Sichuan Basin. *Acta Pet. Sin.* **2021**, *42*, 259–272.
11. Ma, X. "Extreme utilization" development theory of unconventional natural gas. *Pet. Explor. Dev.* **2021**, *48*, 326–336. [CrossRef]

12. Guo, X.S. Controlling factors on shale gas accumulations of Wufeng-Longmaxi Formations in Pingqiao shale gas field in Fuling area, Sichuan Basin. *Nat. Gas Geosci.* **2019**, *30*, 1–10. [CrossRef]
13. Wang, P.W.; Zou, C.; Li, X.J.; Jiang, L.W.; Li, J.J.; Mei, J.; Zhang, Z.; Li, Q.F. Main geological controlling factors of shale gas enrichment and high yield in Zhaotong demonstration area. *Acta Pet. Sin.* **2018**, *39*, 744–753.
14. Nie, H.K.; Chen, Q.; Zhang, G.R.; Sun, C.X.; Wang, P.W.; Lu, Z.Y. An overview of the characteristic of typical Wufeng–Longmaxi shale gas fields in the Sichuan Basin, China. *Nat. Gas Ind. B* **2021**, *8*, 217–230. [CrossRef]
15. He, C.; Wan, Y.J.; Su, Y.H.; Luo, W.J.; Geng, X.Y. Comprehensive Analysis of Influencing Factors for Shale Gas Well Production in Weiyuan Gas Field. In Proceedings of the International Field Exploration and Development Conference 2019, Xi'an, China, 16 October 2019.
16. Chai, Y. Study on the Main Controlling Factors of Shale Gas Production Based on Grey Correlation Method. *J. Chon. Uni. Sci. Tech. Nat. Sci. Ed.* **2018**, *20*, 32–34.
17. Jayakumar, R.; Rai, R. Impact of Uncertainty in Estimation of Shale-Gas-Reservoir and Completion Properties on EUR Forecast and Optimal Development Planning: A Marcellus Case Study. *SPE Res. Eval. Eng.* **2014**, *17*, 60–73. [CrossRef]
18. Chen, X.; Xu, J.L.; Li, J.; Xiao, J.F.; Zhong, S.C. An Analysis of Main Controlling Factors of Production for Horizontal Shale Gas Well in Weiyuan Block. J. Southwest. *Pet. Uni. Sci. Tech. Ed.* **2020**, *41*, 63–74.
19. Jia, C.Y.; Jia, A.L.; He, D.B.; Wei, Y.S.; Qi, D.Y.; Wang, J.L. Key factors influencing shale gas horizontal well production. *Nat. Gas Ind.* **2017**, *37*, 80–88.
20. Zhao, X.; Rui, Z.; Liao, X.; Zhang, R. A simulation method for modified isochronal well testing to determine shale gas well productivity. *J. Nat. Gas Sci. Eng.* **2015**, *27*, 479–485. [CrossRef]
21. Xu, Y.; Liu, Q.; Li, X.; Meng, Z.; Yang, S.; Tan, X. Pressure transient and Blasingame production decline analysis of hydraulic fractured well with induced fractures in composite shale gas reservoirs. *J. Nat. Gas Sci. Eng.* **2021**, *94*, 104058. [CrossRef]
22. Cui, Y.; Jiang, R.; Wang, Q.; Liu, X. Production performance analysis of multi-fractured horizontal well in shale gas reservoir considering space variable and stress-sensitive fractures. *J. Pet. Sci. Eng.* **2021**, *207*, 109071. [CrossRef]
23. Han, D.; Kwon, S. Application of Machine Learning Method of Data-Driven Deep Learning Model to Predict Well Production Rate in the Shale Gas Reservoirs. *Energies* **2021**, *14*, 3629. [CrossRef]
24. Niu, W.; Lu, J.; Sun, Y. A Production Prediction Method for Shale Gas Wells Based on Multiple Regression. *Energies* **2021**, *14*, 1461. [CrossRef]
25. Yin, D.; Wang, D.; Zhang, C.; Duan, Y. Shale Gas Productivity Predicting Model and Analysis of Influence Factor. *Open Pet. Eng. J.* **2015**, *8*, 203–207.
26. Guo, W. Differentiation Analysis on Shale Gas Production of the Horizontal Wells in Sichuan Weiyuan Block. *Sci. Tech. Eng.* **2018**, *18*, 228–233.
27. Deng, J.L. Introduction to the grey system theory. *Inner Mongolia Elec. Power.* **1993**, *3*, 51–52.
28. Fang, B.; Hu, J.; Xu, J.; Zhang, Y. A semi-analytical model for horizontal-well productivity in shale gas reservoirs: Coupling of multi-scale seepage and matrix shrinkage. *J. Pet. Sci. Eng.* **2020**, *195*, 107869. [CrossRef]
29. Tan, X.R.; Deng, J.L. A new method for the multi-variable statistical analysis. *Stat. Res.* **1995**, *3*, 46–48.
30. Liu, Y.; Zhang, X.; Shi, J.; Guo, W.; Kang, L.; Yu, R.; Sun, Y.; Wang, Z.; Pan, M. A reservoir quality evaluation approach for tight sandstone reservoirs based on the gray correlation algorithm: A case study of the Chang 6 layer in the W area of the as oilfield, Ordos Basin. *Energy Explor. Exploit.* **2021**, *39*, 1027–1056. [CrossRef]
31. Zhou, Y.X.; Zhao, A.K.; Yu, Q.; Zhang, D.; Zhang, Q.; Lei, Z.H. A new method for evaluating favorable shale gas exploration areas based on multi-linear regression analysis: A case study of marine shales of Wufeng-Longmaxi Formations, Upper Yangtze Region. Sediment. *Geol. Teth. Geol.* **2021**, *41*, 387–397.
32. Liu, Y.; Lai, F.; Zhang, H.; Tan, Z.; Wang, Y.; Zhao, X.; Tan, X. A novel mineral composition inversion method of deep shale gas reservoir in Western Chongqing. *J. Pet. Sci. Eng.* **2021**, *202*, 108528.
33. Tang, L.; Song, Y.; Li, Q.; Pang, X.; Jiang, Z.; Li, Z.; Tang, X.; Yu, H.; Sun, Y.; Fan, S.; et al. A Quantitative Evaluation of Shale Gas Content in Different Occurrence States of the Longmaxi Formation: A New Insight from Well JY-A in the Fuling Shale Gas Field, Sichuan Basin. *Acta Geol. Sin. Engl. Ed.* **2019**, *93*, 400–419. [CrossRef]
34. Li, C.Y. Analysis of Affecting Factors and Productivity Prediction of Shale Gas Well in ZT Block. Master's Thesis, Xi'an Shiyou University, Xi'an, China, 2020.

Article

Pore Structure in Shale Tested by Low Pressure N_2 Adsorption Experiments: Mechanism, Geological Control and Application

Feng Liang [1,2], Qin Zhang [1,2,*], Bin Lu [1,2], Peng Chen [1,2], Chi Su [3], Yu Zhang [4] and Yu Liu [5,*]

[1] National Energy Shale Gas R&D (Experiment) Center, Langfang 065007, China; liangfeng05@petrochina.com.cn (F.L.); lubin0921@163.com (B.L.); chenpeng52169@petrochina.com.cn (P.C.)
[2] PetroChina Research Institute of Petroleum Exploration & Development, Beijing 100083, China
[3] Institute of Geology and Geophysics, Chinese Academy of Sciences, Beijing 100029, China; suchi@cashq.ac.cn
[4] Key Laboratory for Thermal Science and Power Engineering of Ministry of Education, Tsinghua University, Beijing 100084, China; yu1016@mail.tsinghua.edu.cn
[5] College of Geoscience and Survey Engineering, China University of Mining and Technology, Beijing 100083, China
* Correspondence: zhangqin2169@petrochina.com.cn (Q.Z.); liuyu@cumtb.edu.cn (Y.L.)

Citation: Liang, F.; Zhang, Q.; Lu, B.; Chen, P.; Su, C.; Zhang, Y.; Liu, Y. Pore Structure in Shale Tested by Low Pressure N_2 Adsorption Experiments: Mechanism, Geological Control and Application. *Energies* **2022**, *15*, 4875. https://doi.org/10.3390/en15134875

Academic Editor: Hossein Hamidi

Received: 20 May 2022
Accepted: 22 June 2022
Published: 2 July 2022

Publisher's Note: MDPI stays neutral with regard to jurisdictional claims in published maps and institutional affiliations.

Abstract: The N_2 adsorption experiment is one of the most important methods for characterizing the pore structure of shale, as it covers the major pore size range present in such sediments. The goal of this work is to better understand both the mechanisms and application of low-pressure nitrogen adsorption experiments in pore structure characterization. To achieve this, the N_2 adsorption molecular simulation method, low-pressure N_2 adsorption experiments, total organic carbon (TOC) analysis, X-ray diffraction (XRD) analysis, scanning electron microscopy (SEM), and a total of 196 shale samples from the Wufeng–Longmaxi formations in the Sichuan basin have been employed in this study. Based on the analytical data and the simulations, two parameters, the connectivity index and the large pore volume index, are proposed. These parameters are defined as the connectivity of the pore system and the volume of large nanopores (>10 nm) respectively, and they are calculated based on the N_2 adsorption and desorption isotherms. The experimental results showed that TOC content and clay minerals are the key factors controlling surface area and pore volume. However, in different shale wells and different substrata (divided based on graptolite zonation), the relative influences of TOC content and clay minerals on pore structure differ. In three of the six wells, TOC content is the key factor controlling surface area and pore volume. In contrast, clay minerals in samples from the W202 well are the key factors controlling pore volume, and with an increase in the clay mineral content, the pore volume increases linearly. When the carbonate content exceeds 50%, the pore volume decreases with an increase in carbonate content, and this may be because in the diagenetic process, carbonate cement fills the pores. It is also found that with increasing TOC content the connectivity index increases and SEM images also illustrate that organic pores have better connectivity. Furthermore, the connectivity index increases as quartz content increases. The large pore volume index increases with quartz content from 0 to 40% and decreases as quartz increases from 40% to 100%. By comparing the pore structure of shale in the same substrata of different shale gas wells, it was found that tectonic location significantly affects the surface area and pore volume of shale samples. The shale samples from wells that are located in broad tectonic zones, far from large-scale faults and overpressure zones, have larger pore volumes and surface areas. On the contrary, the shale samples from shale gas wells that are located in the anticline region with strong tectonic extrusion zones or near large-scale faults have relatively low pore volumes and surface areas. By employing large numbers of shale samples and analyzing N_2 adsorption mechanism in shale, this study has expanded the application of N_2 adsorption experiment in shale and clarifies the effects of sedimentary factors and tectonic factors on pore structure.

Keywords: pore structure; shale gas; N_2 adsorption experiment; molecular simulation; pore connectivity

1. Introduction

Shale gas in the USA and China has proven to be enormously successful, and these methane molecules of natural gas are mainly stored in pores of the shale reserve [1,2]. Pore systems in shale are important for gas storage and transport [3–6]. From the start of shale gas development and production, pore structure has been one of the key research topics [7,8], and numerous studies have been carried out on pore systems in shale [9,10]. These have shown that shale contains organic pores and inorganic pores, e.g., within clays or formed by carbonate dissolution [11,12]. The pore sizes in shale range from 0.4 nm to greater than 10,000 nm [13]. The variations in pore types and sizes make pore structure in shale difficult to characterize [14,15]. Many different methods have been applied, and these methods can be divided into quantitative, experimental and analytical methods, such as low-pressure N_2 adsorption (LP-N_2-GA), high-pressure mercury injection (HPMI), low-pressure CO_2 adsorption (LT-CO_2-GA), and nuclear magnetic resonance spectroscopy (NMR), cryoporometry, and visible, qualitative characterization methods, such as scanning electron microscopy (SEM), nano-computerized tomography (nano-CT), and atomic force microscopy (AFM) [16–21]. Each method has its own advantages and disadvantages [22]. LT-CO_2-GA can measure pore sizes in the 0.4–2.0 nm range [14]. HPMI can measure pores in the 3–10,000 nm range, but the experimental pressure needed for shale samples is relatively high, and this may alter the pore system [23]. SAXS and SANS measure the total pore structure including the "accessible" pores (i.e., those pores that can be filled by gas from outside the shale sample) and "inaccessible" pores (those pores that cannot be filled by gas from outside the shale), but SAXS and SANS need further improvement to obtain accurate, accessible pore structure information as researchers usually get the total pore information (including inaccessible pore and accessible pore information) from SAXS and SANS [24]. SEM analysis permits easy identification of the pore type in shale, but the information is qualitative because it is visual, and limited by the sample size [25]. Focused ion beam-scanning electron microscope (FIB-SEM) and nano-CT can provide quantitative pore structure data, such as pore volume and pore size distribution, but these methods cannot see pores smaller than 30 nm [26]. Such pores (pore size < 30 nm) typically contribute more than half of the total pore surface area in shale [26]. LP-N_2-GA provides quantitative data on pores in the 1–300 nm range, and these pores contribute the largest fraction of the total pore volume [14]. Thus, this method is considered to be the most important method to characterize the "accessible" pores in shale.

Knowledge of pore evolution over geological time and factors that influence pore evolution, such as the mineralogy and the TOC content and maturity, are essential for understanding mechanisms of shale gas storage and shale gas accumulation [27,28]. Because of the heterogeneity in shale, understanding controlling factors is problematic. If several shale samples are employed in studies of pore evolution and pore influence factors, it is difficult to distinguish the key controlling factors due to the sample heterogeneity [16]. Some studies have found that the pore surface area and pore volume increase with increasing TOC content in shale [29,30]. However, other researchers reported that there is no obvious correlation between the Brunauer–Emmett–Teller (BET) surface area and the TOC content, and that, furthermore, clay minerals have more significant effects on the BET surface area than the TOC content [31].

The N_2 adsorption experiment is a general method used to characterize mesoporous (2–50 nm) materials [32]. The application of N_2 adsorption experiments to shale is mainly based on the method adopted from material science [33]. Shale is different from artificial mesoporous materials, as the pore sizes of most of artificial materials cover a narrow range. In contrast, the pore system of shale contains not only micropores (<2 nm) but also mesopores (2–50 nm) and macropores (>50 nm) [34,35]. This variability would affect pore structure results if the analytical methods used in material science were applied directly. For the N_2 adsorption experiments, there are Horvath Kavazoe (HK), BET, Barrett Joyner Halenda (BJH) and density functional theory (DFT) models for analyzing the N_2 adsorption experimental data to obtain pore structure data. The BET model is usually used to obtain the surface area of the pore system in shale, and the BJH model is usually used to obtain the pore size distribution and the pore volume [36]. In addition, when studying the pore system of shale, the pore connectivity and gas phase state in the pore space are also important [36]. In order to systematically analyze the pore structure of shale, molecular simulations are used to help analyze the N_2 adsorption/desorption isotherms to obtain the pore structure in shale [37,38].

There is still a need to expand and further study the application of low-pressure N_2 adsorption experiments in characterizing the pore structure in shale. In addition, the key factors controlling pore characteristics in shale are still unclear. To further understand pore systems in shale, 196 organic shale samples from the Wufeng–Longmaxi formations in the Sichuan basin were collected, and N_2 adsorption experiments, XRD experiments, SEM experiments, and TOC tests were performed in this study. Longmaxi shale in the Sichuan Basin is the most successful gas shale in China. Most of the commercial shale gas production of China is provided by the Longmaxi shale [39]. In addition, the molecular simulation method was also employed to investigate the N_2 adsorption mechanism in shale.

2. Materials and Methods

2.1. Samples

A total of 196 shale samples from the Wufeng–Longmaxi formations were collected from six shale gas wells in the Sichuan basin. The units in the formations are subdivided by using graptolites, which are abundant in the Wufeng–Longmaxi. The formations can thus be divided into 13 zonations, and among these, there are 12 zonations in Katian, Hinantian, Rhuddanian, and Aeronian as shown in Figure 1a [40]. Based on the graptolite zonation, lithofacies, sedimentary environment, and natural gamma log (GR), the shale samples are divided into four substrata for shale gas exploration: Wufeng, LM1-3, LM4-5, and LM6-8 (Figure 1b) [41].

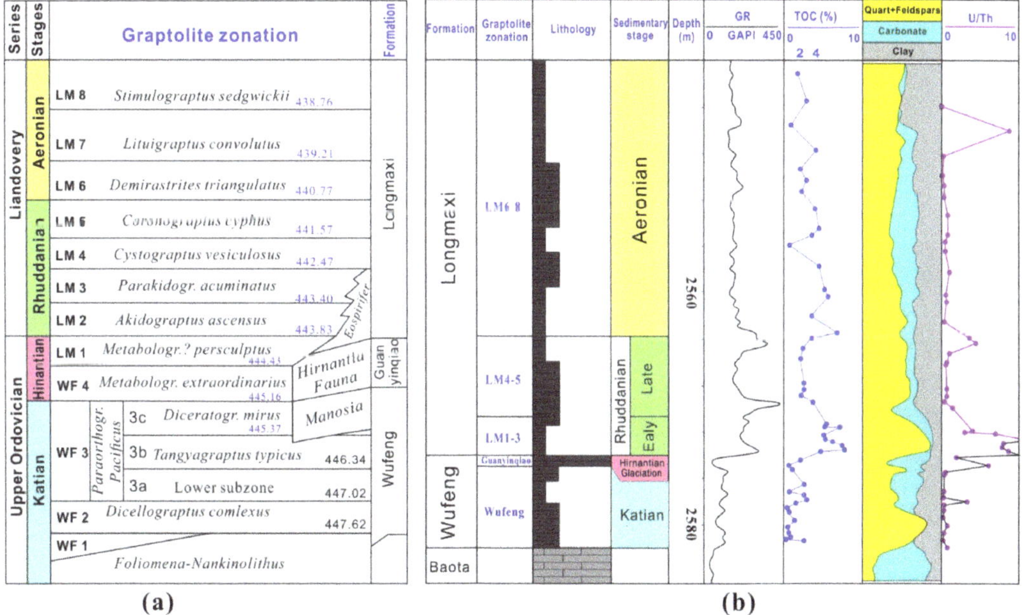

Figure 1. (a) Subdivision of the Wufeng–Longmaxi formations into units based on graptolite zonation: (Katian, Hinantian, Rhuddanian, and Aeronian: WF1-4 and LM1-8) [40]; (b) Typical stratigraphic column for the Wufeng–Longmaxi formations and its subdivisions in the Weiyuan location used in this study [41].

The depth of the Wufeng–Longmaxi formations generally increases from southwest to northeast (Figure 2) [3]. The six wells are mainly in the southwestern Sichuan Basin where the Wufeng–Longmaxi is shallower, with the exception of the B201 well, located more toward the center of the basin. The true vertical depths (TVD) of the Wufeng–Longmaxi shale samples analyzed range from 2500 to 4300 m.

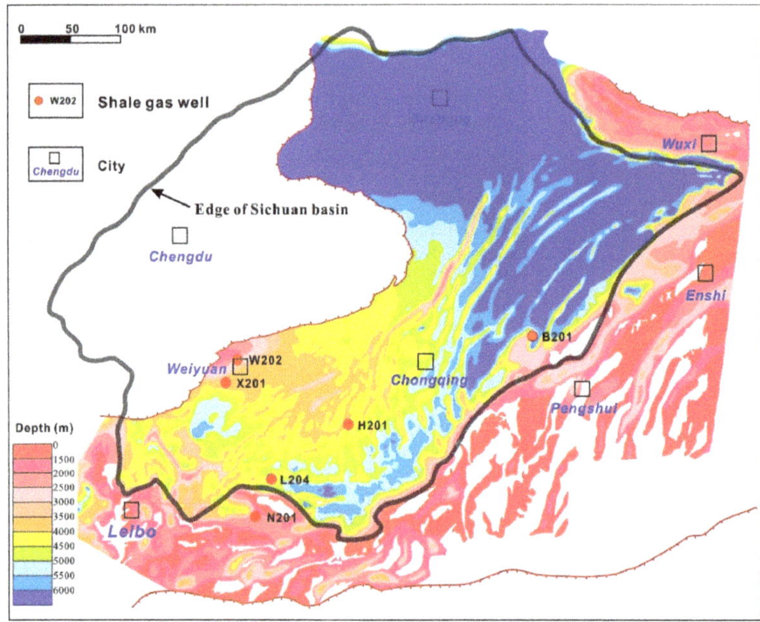

Figure 2. The true vertical depth (TVD) of the Wufeng–Longmaxi formations in the Sichuan basin and the location of the wells [3].

2.2. Methods

2.2.1. Low-Pressure N_2 Adsorption Experiment

LP-N_2-GA experiments were used to characterize the properties of nanopores (1.0–300 nm) in shale samples by employing the Micromeritics ASAP 2420 instrument. First, the shale samples were pulverized to about 60–80 mesh. The 1.5–2.5 g pulverized shale samples were used in the LP-N_2-GA experiment. Then, degassing was performed at an ambient temperature of 22.0 °C before testing N_2 adsorption and desorption amounts. After degassing, the N_2 adsorption amount was tested at a testing temperature of −195.8 °C (77.35 K) with an increase in the P/P_0 from 0.01 to 0.99. The equilibration interval was set to 10 s. During the desorption process, P/P_0 decreased from 0.99 to 0.14 and the N_2 desorption isotherms were obtained.

2.2.2. Simulation Methods

The grand canonical Monte Carlo (GCMC) method was used to simulate the N_2 adsorption in shale, and thus it was an equilibrium calculation rather than a time-dependent calculation. The simulation temperature is also 77.35 K, which is consistent with the experimental temperature. As pores in shale are heterogeneous and complex, different-sized slit pores were constructed to study the N_2 adsorption behavior. As the organic matter (OM) in Longmaxi shale of the Sichuan basin was over-mature [42], the surface of the reconstructed pore was composed of carbon atoms only. In the simulation, the forcefield used was condensed-phase optimized molecular potentials for atomistic simulation studies (COMPASS) [40]. As N_2 adsorption in shale is physical adsorption, the forces between the N_2 molecule and pore model are mainly intermolecular forces (van de Waals force). In the COMPASS forcefield, the van de Waals force are calculated by LJ 9-6 equation (Equation (1)) [43]:

$$E_{Van} = \varepsilon_{ij}[2\left(\frac{r_{ij}^0}{r_{ij}}\right)^9 - 3(\frac{r_{ij}^0}{r_{ij}})^6].$$

(1)

In this work, ε_{ij} and r_{ij}^0 are the van der Waals parameters between N atoms and carbon atoms; r_{ij} represents the distance between the two atoms. More details about the LJ9-6 Equation can found in [43].

2.2.3. TOC Content

The TOC content was measured by using the LECO carbon and sulphur analyzer following the Chinese standard GB/T 19145-2003 (determination of total organic carbon in sedimentary rock). Before the experiment, the shale samples were pulverized to less than 60 mesh. Then, a hydrochloric acid solution was added to a known mass of the powdered samples for 2 h to remove the carbonate minerals. After washing and drying, the residues were weighed and analyzed for their TOC content.

2.2.4. XRD

The XRD analyses were performed by using the RINT-TTR3 X-ray diffractometer following the Chinese petroleum industry standard SY/T 5163-2018 (analytical method for clay minerals and ordinary non-clay minerals in sedimentary rocks by X-ray diffraction). Before the experiment, the shale samples were pulverized to 40 μm. During the analysis, the scan range was 2θ = 5–45 °, and the scan speed was 2 °/min.

3. Results

3.1. TOC Data

Figure 3 illustrates the average TOC contents of shale samples from each substratum in the six shale gas wells. It can be seen that the average TOC contents of the Wufeng substrata in the six shale gas wells vary greatly, ranging from 1.1% to 7.3%. In contrast, the average TOC contents of LM1-3 substrata are all very high, ranging from 4.6% to 6.0%. The average TOC contents of LM4-5 and LM6-8 substrata are much lower than those of LM1-3 substrata. The average TOC contents of LM4-5 and LM6-8 substrata range from 2.3% to 3.4% and 2.6% to 3.4%, respectively.

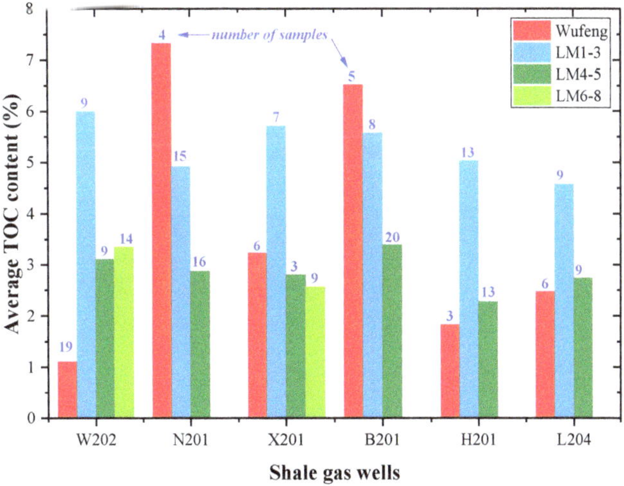

Figure 3. The average TOC contents of shale samples from each substratum in the six shale gas wells.

3.2. Mineralogy

The mineral components of shale samples are determined by the XRD results (Figure 4). The shale samples from different substrata and different wells are mainly composed by clay, quartz, potash feldspar, plagioclase, calcite, dolomite, pyrite, and other minerals (Figure 4). Among these mineral types, the contents of clays, quartz, and carbonates (calcite and dolomite) are usually larger by 20% each. Comparably, the content of other minerals such as pyrite is mainly less than 5%. By comparing the mineral composition of different shale samples from different shale wells and different substrata, it can be found that the mineral composition of different substrata differ significantly. However, the shale samples with the same substrata in different shale gas wells often have a similar mineral composition. The shale samples of LM1-3 substrata usually have the lowest clay mineral and highest quartz contents of all substrata in each well (Figure 4). This is mainly because the mineral composition is mainly determined by the sedimentary environment.

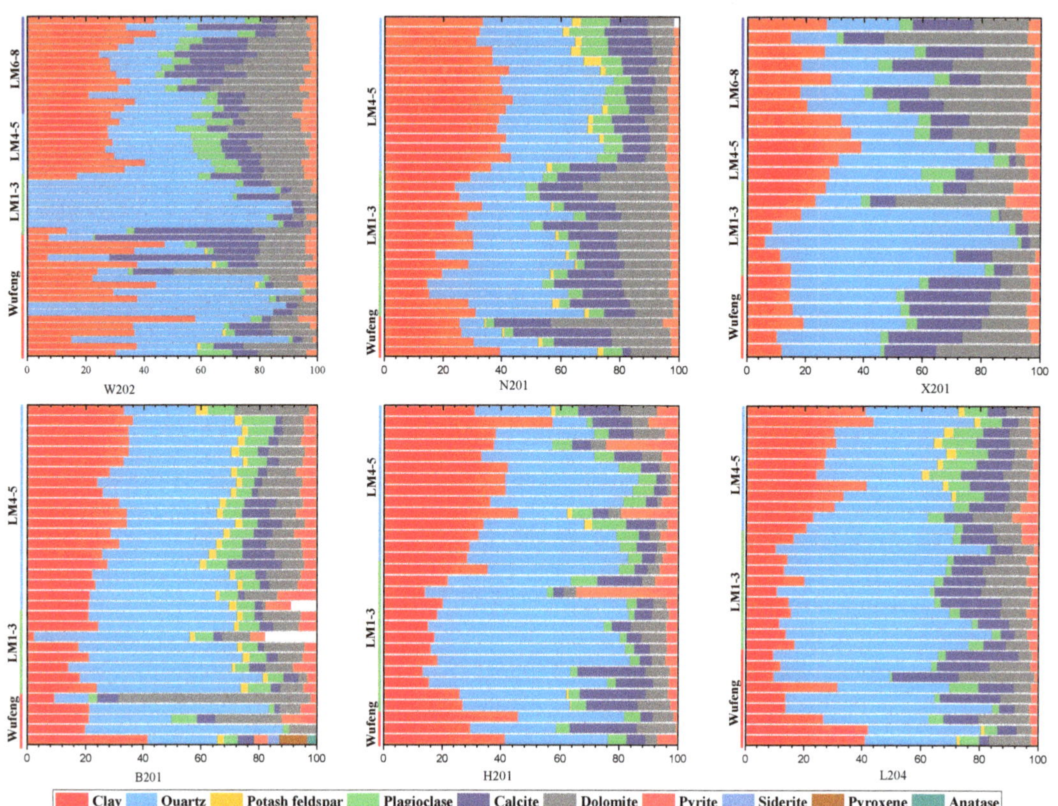

Figure 4. The mineral composition of the shale samples from the six wells.

3.3. Simulation Results and the Mechanism of N_2 Adsorption in Shale

Shales have a pore size that ranges from <1 nm to >1000 nm. Thus, different-sized slit-shaped pores were used in the model. Figure 5a shows the N_2 adsorption isotherms for different-sized slit pores at 77.35 K. It can be seen that the N_2 adsorption behavior differs significantly in different pore sizes. In 1-nm pores, the N_2 adsorption volume reaches the largest value at low pressure and then remains flat with increasing pressure. In 2-nm slit pores, the N_2 adsorption amount reaches the maximum value at 8 kPa, and then remains stable with an increase in pressure (Figure 5a). It should be noted that in small nanoscale pores, the N_2 molecule fills up the pore space at the low-pressure stage. In these small pores, there is little space for N_2 storage, and thus, with increasing pressure, the volume of N_2 does not increase. In larger pores, the pressure at which the maximum volume of N_2 fills the pore space is greater as shown in Figure 5a. For example, in 2-nm pores, the pressure at which the pore space is filled is about 8 kPa. Comparably, in 3-nm, 4-nm and 6-nm slit pores, the corresponding pressure (the pressure at which the pore space is filled) is 33 kPa, 57 kPa, and 96 kPa, respectively (Figure 5a).

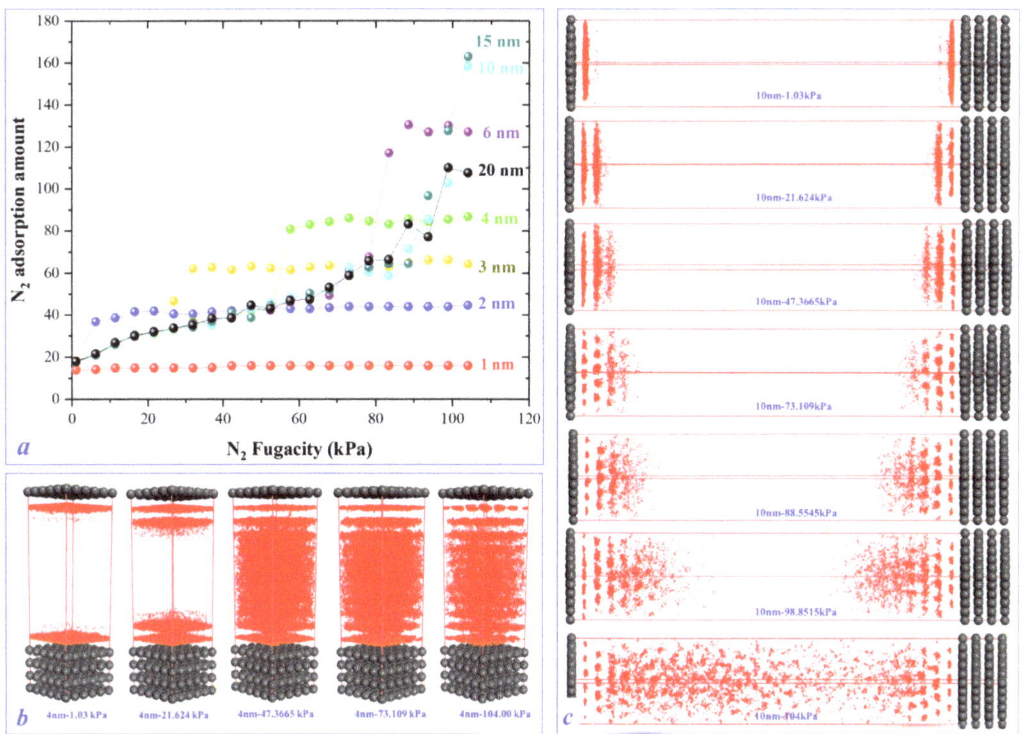

Figure 5. (a) N_2 adsorption isotherms in different-sized slit pores; (b) N_2 density distribution in the 4 nm slit pore; (c) N_2 density distribution in 10 nm slit pore.

From Figure 5a, it can also be seen that small nanoscale pores have their own dominant pressure stages. At 8 kPa–20 kPa, the 2-nm pores contain the largest volume of N_2. At 35 kPa–50 kPa, the 3-nm pores contain the largest volume of N_2 and at 60 kPa–80 kPa, the largest volume of N_2 is in the 4-nm pores. These dominant pressure stages for each size of slit pores can be used to identify the pore structure based on low-pressure N_2 adsorption experiments. Figure 5b shows the N_2 density distribution in 4-nm pores at different pressures. By increasing the pressure, N_2 adsorption is performed layer by layer. At 1.03 kPa, there is a main adsorption layer in 4-nm slit pores. At 21.6 kPa and 47.3 kPa, there are two and three main adsorption layers, respectively. At 73.1 kPa, N_2 would fill up the whole pore space.

Figure 5c shows the N_2 volume density distribution in 10nm pores at different pressures. When the pressure is less than 47.3 kPa, N_2 adsorption in 10-nm pores is similar to that in 4-nm pores. However, when the pressure is greater than 73.1 kPa, the N_2 adsorption behavior in 10-nm pores is quite different. In 10-nm pores, there is more space, and more layers of N_2 can be formed at relatively high pressures, as shown in Figure 5. By analyzing Figure 5a–c, it can be also seen that, in different-sized slit pores, the N_2 saturation pressures are different. This makes the experimental N_2 adsorption and desorption isotherms difficult to analyze.

3.4. Experimental Adsorption and Desorption Isotherms

Considering that there are too many experimental data to put all the N_2 adsorption and desorption data of 196 samples in one or two figures, the adsorption and desorption isotherms of the shale samples from well B201 were chosen to represent the adsorption and desorption characteristics as shown in Figure 6. It can be seen that when the relative pressure (P/P0) = 0.01, the initial mass of N_2 adsorbed by the shale samples is greater than $4 \text{ cm}^3/\text{g}$, indicating that some shale samples have relatively large surface area or micropore volume (Figure 6). When P/P0 increases from 0.1 to 0.8, the volume of N_2 adsorbed by all the shale samples increases slowly.

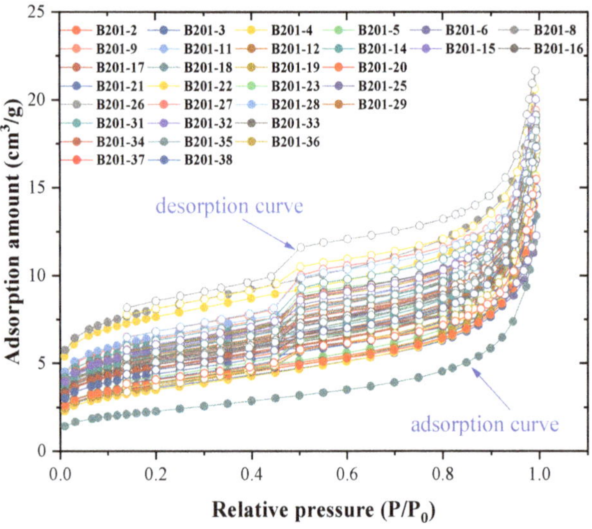

Figure 6. The experimental N_2 adsorption and desorption curves of shale samples from B201 well.

The adsorption isotherms, desorption isotherms and hysteresis loops of the gas adsorption reflect the pore structure of the porous material [44,45]. Based on molecular simulation, the adsorption and desorption isotherms of individual pores can be obtained [44] (Figure 7). As the shale contains different-sized and different-shaped pores, by analyzing the adsorption and desorption isotherms of different pores, the experimental adsorption and desorption isotherms can be better understood (Figure 7). Based on the simulation results in the work of Neimark et al. [45], with increasing pore size, the hysteresis loops shift to greater pressure (condensation and evaporation pressure both increase) (Figure 7). However, when we analyze desorption isotherms of all the shale samples, it can be found that the evaporation pressures are all around 0.5 (P/P0). It indicates that there are pores around 6 nm that provide significant pore volume.

By analyzing the N_2 adsorption and desorption isotherms of 196 shale samples, there are three types of N_2 adsorption desorption isotherms. We chose a representative curve for each type as shown in Figure 8. Among the three N_2 adsorption and desorption isotherms, N201-4 has the largest adsorption volume at each pressure. The difference between the adsorption and desorption loop of shale sample N201-4 is also the largest. In contrast, the adsorption and desorption loop of shale sample N201-37 is quite narrow and the N_2 adsorption amount is less than the other two. As shown in Figure 8, when the relative pressure is greater than 0.9, the adsorption amounts increase sharply for all shale samples. Based on Figures 5 and 7, with increasing pressure from 0.9 to 1.0, N_2 condensation occurs in all large pores (>10 nm). Before that, smaller pores are fully filled with molecular N_2. And the surface of the large pores has been covered by the N_2 molecules. Thus, the incremental adsorption amounts are from the inner space of the large pores. These spaces are the storing places for free gas.

Figure 7. Adsorption and desorption loops in different sized and different shaped ideal pores [37,38,45]. Adapted with permission from Ref. [37]. Copyright 2016 Elsevier Ltd. Adapted with permission from Ref. [38]. Copyright 2015 Elsevier Inc. Adapted with permission from Ref. [45]. Copyright 2000 American Physical Society.

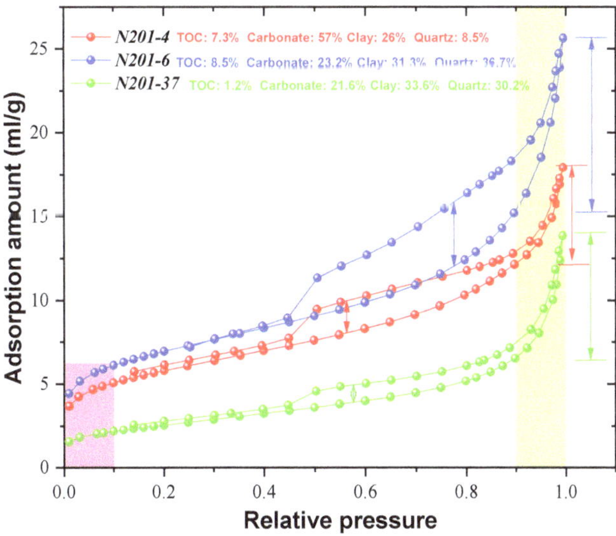

Figure 8. Representative adsorption and desorption isotherms among all the experimental N_2 adsorption data.

For shale gas studies, the connectivity of pore system and the volume of large nanopores (>10 nm, having large storage capacity and gas circulation) is of significant interest. From Figure 7, it can be seen that the adsorption and desorption curves of the open pores which have good connection in the pore system have wide hysteresis loops, whereas those of closed-end pores have quite narrow hysteresis loops. Thus, the hysteresis loops could reflect the connectivity of the pore system. When the P/P_0 is smaller than 0.5, the adsorption curve overlaps the desorption curve. When the P/P_0 is larger than 0.8, the differences between the adsorption amounts on the adsorption curve and desorption curve decrease rapidly. The differences between the adsorption amounts on the adsorption curve and desorption curve at $P/P_0 = 0.8$ and $P/P_0 = 0.5$ could represent the N_2 hysteresis loops. In addition, if there are many large-sized pores (>6 nm), the adsorption amount would significantly increase when the $P/P0 > 0.9$ (Figures 5 and 7). At this pressure stage, N_2 condensation begins and the adsorption amount is controlled by the volume of shale pore. These pores are very significant for the free gas storage. Based on the mechanism of N_2 adsorption in nanopores and the experimental N_2 adsorption and desorption isotherms of the shale samples (Figures 5 and 7), we propose two parameters to reflect (i) the connectivity of the pore volume and (ii) the volume of large nanopores in shale, namely the connectivity index and large-pore volume index. The calculation methods are listed in Equations (2) and (3). The larger connectivity index means a larger hysteresis loop and better connectivity, whereas the larger pore volume index means there is more volume in large pores for N_2 condensation, which also means more free gas can be stored in these shales:

$$\text{Connectivity Index} = \text{De}_{p/p0} = 0.8 - \text{AD}_{p/p0} = 0.8 + \text{De}_{p/p0} = 0.5 - \text{AD}_{p/p0} = 0.5 \quad (2)$$

$$\text{Large-pore volume index} = \text{AD}_{p/p0} = 1.0 - \text{AD}_{p/p0} = 0.9, \quad (3)$$

where: $\text{De}_{p/p0 = 0.8}$ is the adsorption amount at $P/P_0 = 0.8$ during the desorption process, $\text{AD}_{p/p0 = 0.8}$ is the adsorption amount at $P/P_0 = 0.8$ during the adsorption process, $\text{De}_{p/p0 = 0.5}$ is the adsorption amount at $P/P_0 = 0.5$ during the desorption process, $\text{AD}_{p/p0 = 0.5}$ is the adsorption amount at $P/P_0 = 0.5$ during the adsorption process, $\text{AD}_{p/p0 = 1.0}$ is the adsorption amount at $P/P_0 = 1.0$ during the adsorption process, and $\text{AD}_{p/p0 = 0.9}$ is the adsorption amount at $P/P_0 = 0.9$ during the adsorption process.

4. Discussion

4.1. Overall Analysis of Factors That Influence Pore Structure

To better analyze the controlling factors of pore structure in organic shale, the BET surface area data of all 196 shale samples are presented in Figure 9a–c. Previous studies suggest that OM pores are the most important components of pore systems in organic rich shale; hence TOC content usually has a good positive correlation with BET surface area [7,46–48]. However, there is no correlation between the BET surface area and TOC content in this study (Figure 9a). In fact, this relationship is very complicated and is caused by several factors, such as tectonic compaction, filling and blocking of primary kerogen pores, and limited thermal conversion into liquid hydrocarbons [49,50]. Therefore, the effect of organic matter on the pore structure of shale samples from different wells may be quite different.

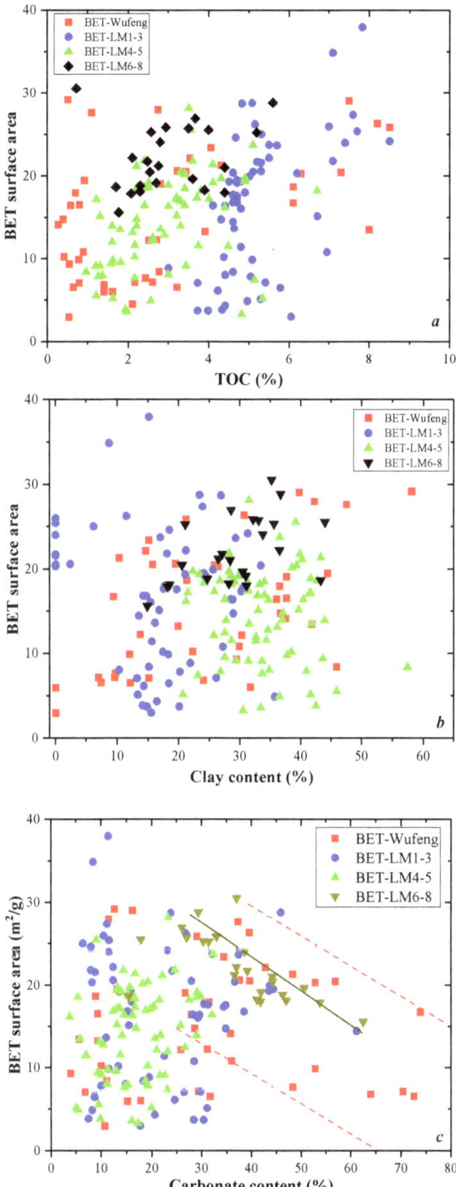

Figure 9. The correlations between BET surface area and TOC content (**a**), clay content (**b**), and carbonate content (**c**) of all 196 shale samples (sample L204-4 has an extremely large BET value, and is not presented in these figures. Data for sample L204-5: TOC-5.25%, quartz content: 65.8%; clay content-11.6%; BET-56.87 m^2/g, BJH volume-0.0534 cm^3/g).

The absence of significant correlations between BET surface area and the clay content of 196 shale samples can be seen in Figure 9b. Yang and Aplin [51] also observed that no apparent correlation between the clay content and specific surface area in the Horn River Group shale in Canada. Some studies have even documented that a slightly negative correlation between the specific surface area and total clay mineral content of some marine shale [52,53]. Chalmers and Bustin [54] attributed these insignificant or negative relationships to the fact that the clay-associated pores are usually mesopores and macropores, which only contribute very limited pore volume and specific surface area. In addition, due to the ductility of clay minerals and the lack of shelter from rigid matrix minerals, tectonic compaction can easily destroy the internal porosity in the shale reservoir with high clay mineral content, leading to a sharp reduction of the pore volume within the shale matrix.

The positive correlation between porosity and quartz content has been reported in studies of major marine shale in south China [55] and Devonian gas shale in northern Canada [56]. Many studies attribute this correlation to rigid matrix minerals with high compressive strength, a high Young's modulus, and a low Poisson's ratio, which can create stress shadows that shield adjacent nanopores, especially the pores related to non-load-bearing minerals from compaction [48,57]. However, this relationship is not significant in this study, even when the carbonate content exceeds 40% and the pore surface area decreases with increasing carbonate content. In fact, carbonate minerals could shelter the pores from compaction, but the excessive carbonate may lead to the decrease of macropores due to precipitation within pore spaces [58].

4.2. Geological Controls on Pore Structure in Shale Samples from Each Well

For shale gas reservoirs from different structural locations, different diagenesis and tectonism can not only directly affect the pore structure, but also indirectly affect the pore structure by affecting other factors that can control the pore structure [30,59,60]. In addition, for the different sedimentary substrata in the Wufeng–Longmaxi formation, the sources of OM and sedimentary microfacies are different, and this results in differences in TOC content and mineral composition (Figures 3 and 4). To help elucidate the factors that influence pore structure in Wufeng–Longmaxi shale, data for samples from different shale gas wells have been analyzed in detail in the following sections. As a result, the six shale gas wells can be divided into three types according to the key factors controlling the pore properties. For each type of shale gas well, the controlling factors of pore structure in organic shale are different (Type I: pore properties are mainly controlled by the TOC content; Type II: pore properties of shales in different substrata are controlled by different factors; Type III: pore properties are not obviously controlled by either TOC content or mineral content.).

4.2.1. Type I: Pore Properties Are Mainly Controlled by the TOC Content

For the shale samples from the N-201 well, the X-201 well, and the B-201 well, there is a strong correlation between the BET surface area and the TOC content as shown in Figure 10a–c. It should be also noted that the strong correlation occurs at whole well scale, which means that for different shale substrata, the correlation is similar. It means the pore BET surface area provided by the organic matter in different shale substrata is nearly the same. The correlation between BJH pore volume and TOC content shows different behaviors as shown in Figure 11a–c. Figure 11a illustrates the relationship between BJH pore volume and TOC content for well N201. For the samples from a given substratum, the BJH pore volume increases with increasing TOC content. However, when comparing the BJH pore volume of shale samples from different substrata, the trends are offset for each substratum. For any given TOC content, the BJH pore volume in samples from LM4-5 substrata is larger than that in shales from LM1-3 and the Wufeng formation. However, the correlation line of LM4-5 substrata is parallel to that of LM1-3 substrata, indicating the increase of organic matter per unit mass in LM4-5 shale provides similar porosities with that in LM1-3 shale. The differences of BJH pore volume between LM1-3 shale and LM4-5 shale are caused by other factors. By comparing the mineral composition of LM1-3 shale and LM4-5 shale in N201 well, it can be seen that the clay mineral of LM4-5 shale is obviously larger than that of LM1-3 shale (Figure 4), indicating clay minerals provide initial pore volume. Based on the initial pores provided by clay, the pore volume of shale increases with the increase of organic matter content, and in different substrata, the difference in clay minerals brings the difference in the initial pore volume (Figure 11b).

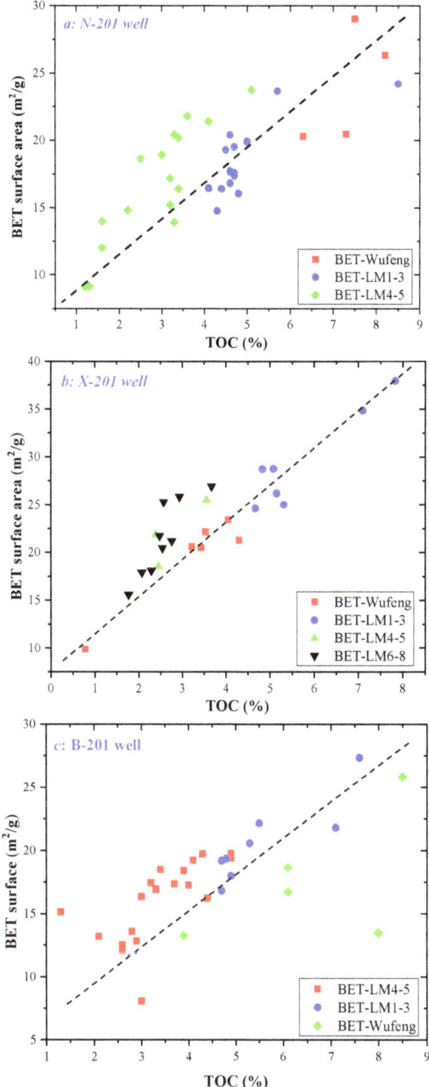

Figure 10. The effect of TOC content on BET surface area the shale samples from (**a**) the N201 well, (**b**) the X201 well, and (**c**) the B-201 well.

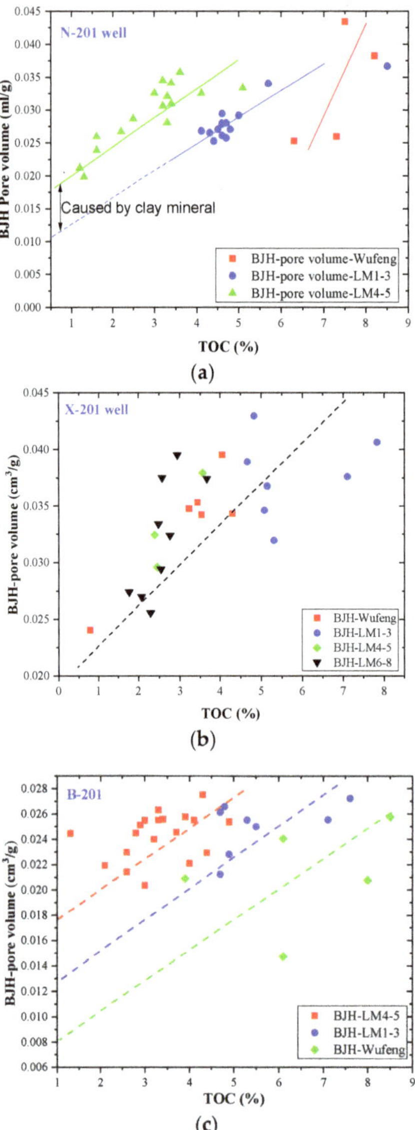

Figure 11. The correlations between BJH pore volume and TOC content of the shale samples from (**a**) the N201 well, (**b**) the X201 well, and (**c**) the B-201 well.

By comparing Figures 10b and 11b, it can be seen that in all the samples for the X201 well, the BET surface area and the BJH pore volume both increase with increasing TOC content. The shales of different substrata in X201 well have similar pore size distributions. This is different from the N201 well. The relationship between BJH pore volume and TOC content in well B201 is more complex (Figure 11c). Although the BET surface increases with increasing TOC content, the overall correlation between BJH pore volume and TOC is quite weak. By analyzing the different substrata separately, the BJH pore volume also increases with increasing TOC content. That means small nanopores were provided by OM and there are many large inorganic pores in shale from the B201 well.

It is important to understand why the BJH pore volume of LM4-5 shale is greater than that of LM1-3 shale with the same TOC content in N201 and B201 well. By comparing the nanopore structures of shales in different substrata in the same well, the effects of sedimentary environment, mineral composition and OM type can be established, as the shales in different substrata in the same well have similar maturity and have experienced a similar tectonic history. LM1-3 substrata have higher TOC values (Figure 3), lower clay content, and higher carbonate content (Figure 4) compared to other substrata. The sedimentary environment determines the OM type, TOC content, quartz, clay, and other mineral contents. The fact that BET surface area increases with increasing TOC for all the Wufeng–Longmaxi formation also indicates that small nanopores characterized by N_2 adsorption are similar in different substrata. However, the larger nanopores characterized by N_2 adsorption are quite different in different substrata. In addition, the clay content of shale samples in LM1-3 substrata is lower than that of LM4-5 substrata shale samples in (Figure 4). This may result in a lower pore volume in LM1-3 shale than that of LM4-5 shale with the same TOC content. Considering that the content of organic matter is much lower than that of clay, in the N201 and B201 wells, organic matter is the main controlling factor, but clay provides some initial pore volume.

4.2.2. Type II: Pore Properties of Shales in Different Substrata Are Controlled by Different Factors

Figure 12a,b illustrate the correction between BET surface area and TOC content and clay content in shale from the W202 well. A significant correlation between the BET surface area and TOC content appears in LM1-3, but not in the Wufeng Formation and LM6-8 (Figure 12a). Milliken et al. [30,59,60] also observed that there is not always a positive correlation between BET surface area and TOC content, because high TOC content could reduce the brittleness of shale and lead to the collapse of some OM pores. In addition, a good positive correlation between BET surface area and clay content appears in the Wufeng Formation shale samples, which is closely related to the high clay mineral content of these samples (Figure 4). It also implies the pore surface area is mainly controlled by clay minerals rather than organic matter (Figure 12b). Ji et al. [61] has previously confirmed that clay-related micro-mesopores can provide a large specific surface area, but the contribution of different clay minerals to a specific surface area varies. In addition, the BET surface area of LM4-5 and LM6-8 substrata do not correlate with individual parameters such as TOC content, clay minerals, and other major mineral contents. In fact, the pore structure of shale reservoir is affected not only by material composition, but also by external conditions [60,62]. Therefore, we suggested that this insignificant correlation of LM4-5 and LM6-8 substrata is related to external factors.

Interestingly, the geological controls of BJH pore volume are a little different from those of BET surface area (Figures 12 and 13). The correlation between pore volume and clay in most samples from the W202 well is stronger than that between pore volume and TOC content, which implies that clay content is the most important factor for BJH pore volume (Figure 13). Significantly, the above relationship does not appear in the LM1-3 substrata. This is mainly because the clay content in LM1-3 shales are quite low, and the TOC content in LM1-3 substrata is also greater than that in other substrata (Figure 3). Organic matter is still one of the main controllers of pore properties (Figure 13b).

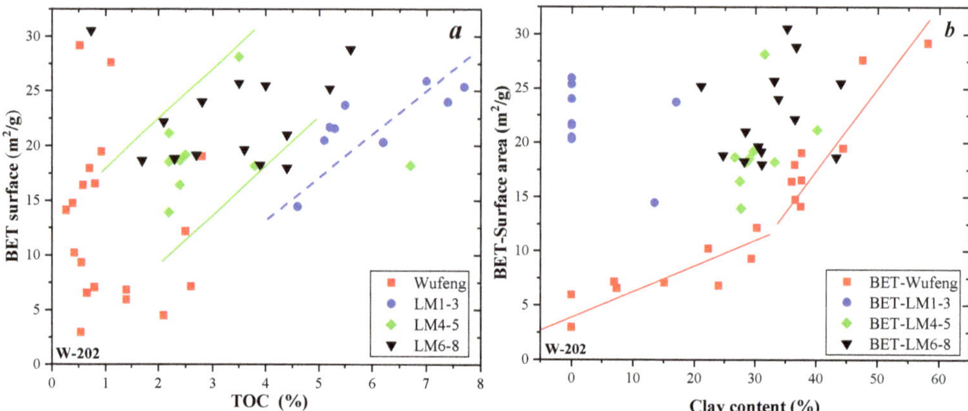

Figure 12. The correlation between the BET surface area and TOC content (**a**), clay mineral content (**b**) in the W202 well.

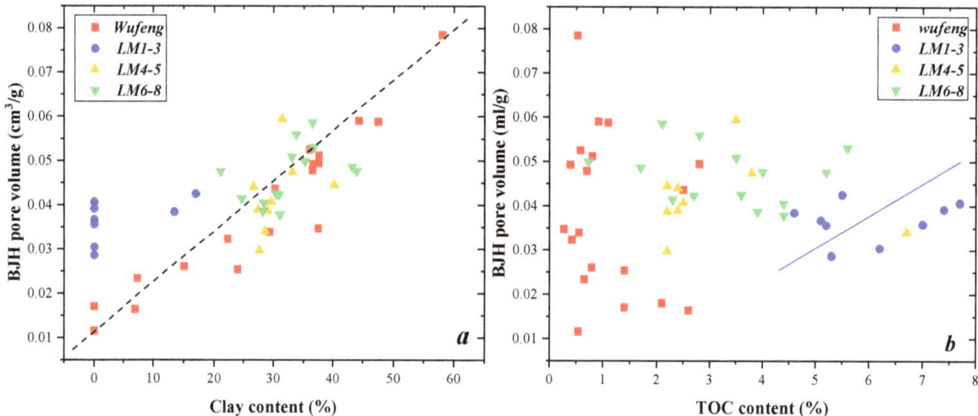

Figure 13. The correlation between BJH pore volume and clay content (**a**), TOC content (**b**) in the W202 well.

Figure 14a,b illustrates the correlation between the BET surface area and the TOC content, clay mineral content, respectively, for well L204. In different substrata, the major control factors are different. For the shale samples in LM1-3 substrata, the BET surface area increases with increasing TOC content. In contrast, for the shale samples in LM4-5 substrata, the BET surface area increases with increasing clay mineral content. In other words, the surface area of LM1-3 shale is mainly controlled by TOC content and that of LM4-5 shale is mainly controlled by clay mineral content. The different controlling factors for LM4-5 shale, LM1-3 shale, and Wufeng formation shale can both be caused by sedimentary environment and tectonic compaction reactions. The differences in sedimentary environment will accompany the difference in mineral composition. In addition, as the contents of brittle mineral, such as quartz, in different substrata are different, the tectonic compaction reactions of the pore system are different, which has been proven by previous studies [62,63].

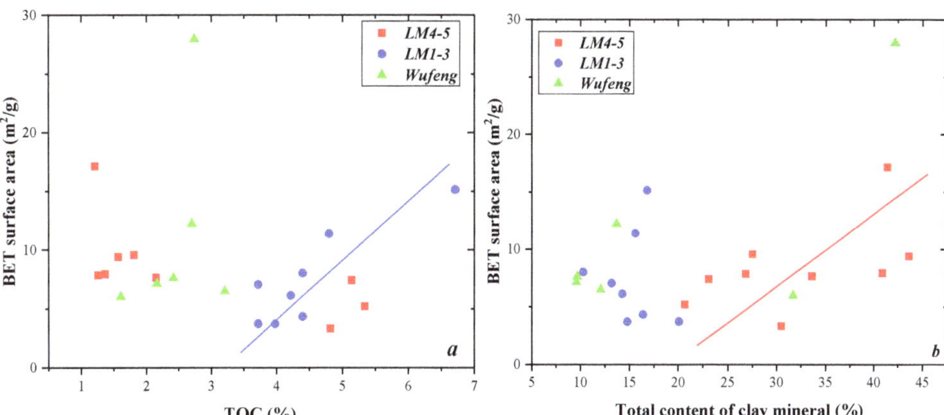

Figure 14. The correlation between BET surface area and the TOC content (**a**), clay mineral content (**b**) in well L204 (sample L204-4 has extremely large BET value, and was not included in the two figures). Information of sample L204-5: TOC-5.25%; quartz content: 65.8%; clay content: 11.6%; BET: 56.87 m^2/g; BJH volume: 0.0534 cm^3/g.

4.2.3. Type III: Pore Properties Are Not Obviously Controlled by either TOC Content or Mineral Content

The pore properties of shale in the H201 well are not obviously controlled by TOC or mineral contents. Figure 15 shows that the BET surface area has no correlation with TOC content, clay content, and carbonate content in different substrata. Although the TOC content of shale samples from LM1-3 substrata is much greater than that of shale samples from LM4-5 substrata, the BET surface area is similar in both cases. In fact, many factors may complicate the relationship between the BET specific surface area and TOC content. For example, violent tectonic compaction in this deformation belt also enhances the robust complexity and heterogeneity of the pore system, resulting in the weakening of the correlation between TOC and parameters [63]. Similarly, although the clay content of shale samples of LM4-5 substrata is much greater than that of shale samples of LM1-3 substrata, the BET surface area is similar in both cases. Some previous studies suggested that due to the ductility of clay minerals and the lack of shelter from rigid matrix minerals, tectonic compaction can easily destroy the internal porosity in the shale reservoir with high clay mineral content, which also leads to the complex relationship between pore structure and clay minerals [54].

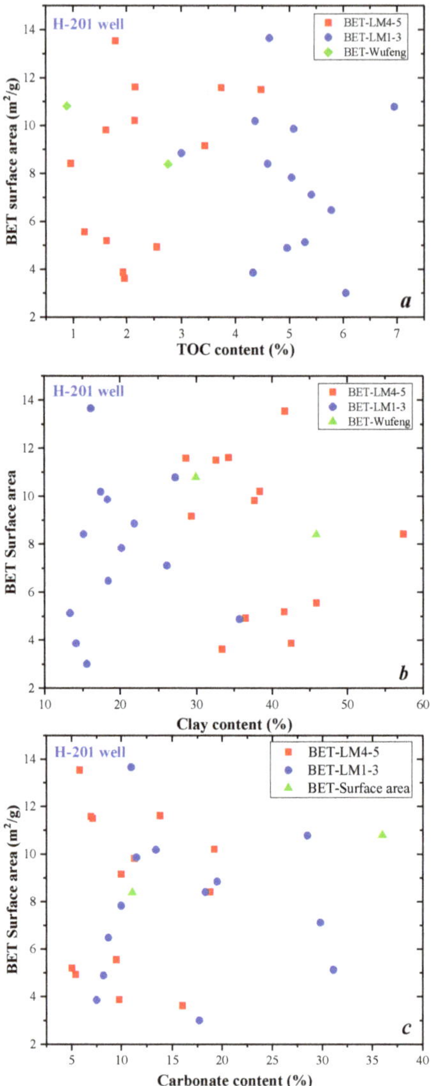

Figure 15. The correlation between BET surface area and TOC content (**a**), clay mineral content (**b**), and carbonate content (**c**) in the H201 well.

4.3. Connectivity of Pores in Shale

Shale pore connectivity is of great significance for hydrocarbons' migration, enrichment, and production behavior [64]. Figure 16 shows the correlation between the connectivity index and the TOC content. A significant correlation between TOC content and connectivity index appears in shale from the W202, N201, X201, and B201 wells, which implies that organic pores have a very significant effect on pore connectivity. In addition, there is no correlation between the TOC content and pore connectivity index of samples in wells H201 and L204, and the connectivity index is quite small. Therefore, we speculated that the pore connectivity of these samples is affected by other factors, such as clay minerals. Higher TOC content means more organic pores, and thus from Figure 16 it can be concluded that organic pores have larger hysteresis loops and better connectivity.

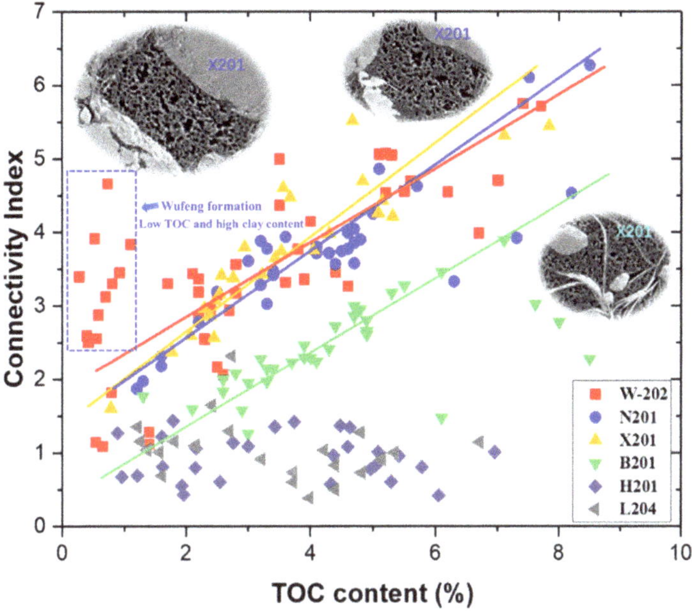

Figure 16. The influence on TOC content on the connectivity index.

Figure 17 illustrates the correlations between the connectivity index and quartz content in wells N201, X201, and B201. In these wells, increasing quartz content correlates with an increase in the connectivity index. This suggests that quartz can facilitate the increase in pore connectivity, presumably because quartz particles form rigid frames that create connected pore networks [65,66].

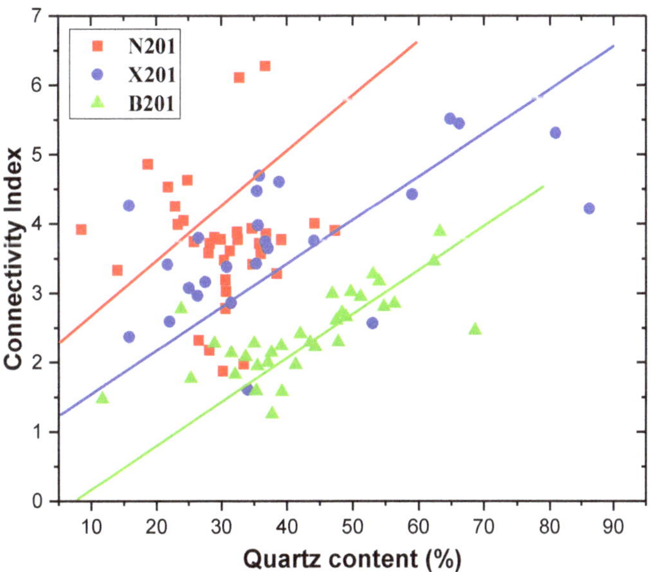

Figure 17. Quartz effect on connectivity index.

As mentioned in Section 3.2, the large pore index reflects internal volumes of pores greater than 6 nm. For these pores, increasing TOC content does not increase the large pore index. In contrast, in the W202, X201, and B201 wells, the large pore index increases with quartz content as quartz increases from 0 to 40% and decreases as quartz increases from 40% to 100% (Figure 18). This suggests that the quartz particles have a significant effect on the large pore index. When quartz content is less than 40%, more quartz particles are conducive to the formation of large pores and provide more space for free gas storage and transport. However, when the quartz content is greater than 40%, the decrease in the large pore index may be due to compression of the pore volume under the horizontal and vertical pressures.

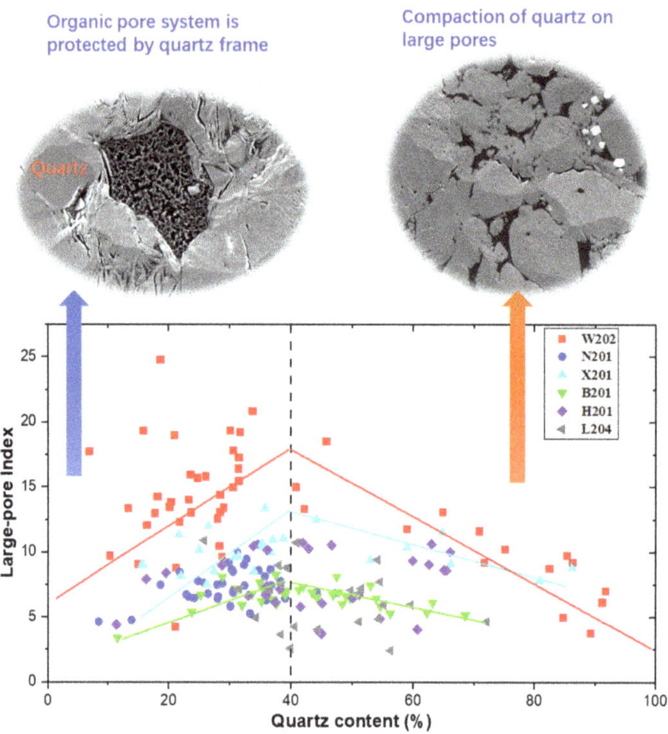

Figure 18. The correlation between the large pore index and the quartz content.

4.4. Factors That Influence Pore Structure in Different Wells

Figure 19 illustrates the pore surface area of LM1-3 samples in different wells. In most wells, the BET surface area tends to increase linearly with increasing TOC content. However, for these LM1-3 substrata shale samples, the correlations between BET surface area and TOC content are different in each well (Figure 19c). The pore surface area of shale from the X201 and N201 wells is greater than that from other samples (Figure 19c). In contrast, the pore surface area of shale samples from L204 well is less than that from other shale wells. The BET surface area and BJH pore volume of LM1-3 shales in the L204 and H201 wells are less than those of other wells. By systemically analyzing the depth, structural form, pore pressure, and tectonic environment, it was found the shale wells of W202, N201, and X201, which have relatively large pore volume and pore surface area, are located in the broad tectonic and overpressure zone (the pressure coefficients are 1.4, 2.03, and 1.95), and these shales are far from large-scale faults. Comparably, the L204 and H201 wells are near faults, and the pore volume and surface area is relatively low (Figure 19). In addition, the B201 is located in an anticline region with strong tectonic extrusion. Strong tectonic compaction will significantly reduce pore volume and specific surface area [62,63,66]. On the contrary, the overpressure and broad structural zones far away from faults are conducive to the preservation of pores.

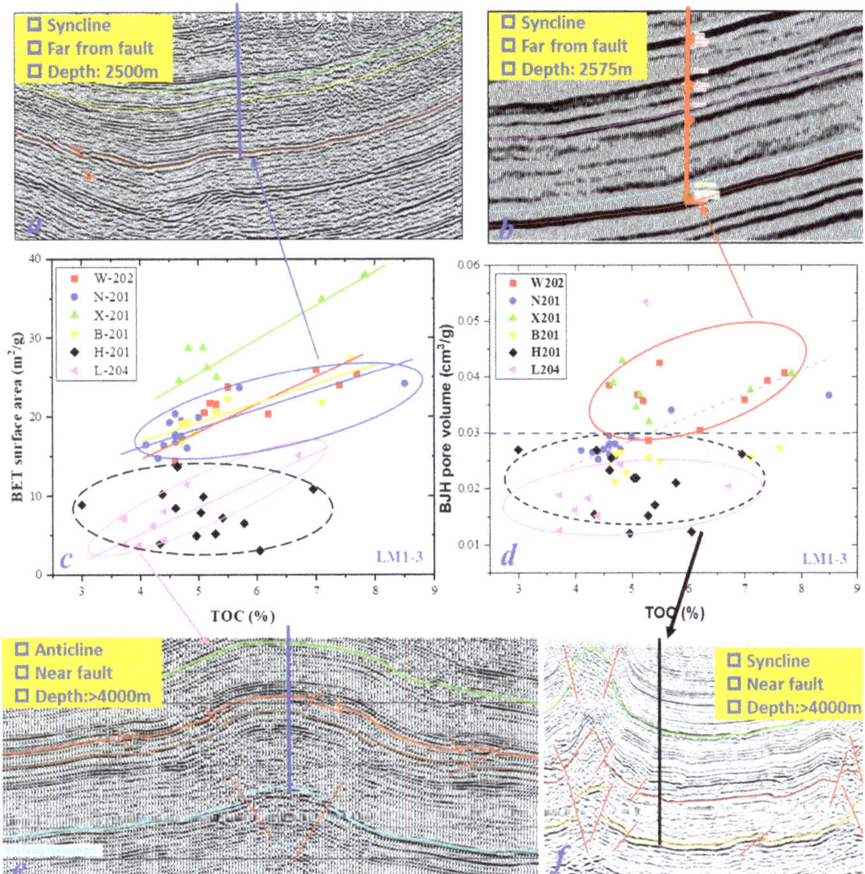

Figure 19. The geological control factors of pore volume and surface area in different shale wells. (**a,b,e,f**): seismic cross-sections of well N201, W202, L201 and H201 wells, (**c,d**). The correlation between TOC content and BET surface area (**c**), BJH pore volume (**d**) (LM1-3 substrata).

5. Conclusions

1. N_2 adsorption and desorption curves are determined by the pore size, pore shapes, and pore connections, and can also provide information on the pore structure and pore connections. Two parameters (connectivity index and large pore volume index) were proposed to reflect the connectivity of pore volume and volume of large nanopores in shale. A larger connectivity index means a larger hysteresis loop and better connectivity, and a larger pore volume index means there is more space volume in large pores for N_2 condensation.

2. In different shale gas wells, the geological control factors are different. OM and clay minerals can both be the key factor of the pore structure. In this study, in the N201, B201, and X201 wells, organic matter is the key factor. With TOC content increasing, the pore volume and surface area both increase linearly. In the W202 shale gas well,

clay minerals are the key factor for pore volume, and with increasing clay mineral content, the BJH pore volume increases linearly.

3. The connectivity of the shale pore system is controlled by organic matter and quartz. With TOC content and quartz content increasing, the connectivity index increases. The large pore index increases with quartz increases from 0 to 40% and decreases with quartz increases from 40% to 100%.

4. Overpressure and broad structural zones, which are also far from faults, are conducive to the preservation of pores. In contrast, shale from wells that are located in anticline regions with strong tectonic extrusion zones or near large-scale faults have relatively low pore volume and surface area.

Author Contributions: Conceptualization, Q.Z. and Y.L.; Investigation, F.L., Q.Z., B.L., P.C., C.S., Y.Z. and Y.L.; Methodology, F.L., Q.Z., B.L., P.C., Y.Z. and Y.L.; Software, Y.Z.; Visualization, F.L. and C.S.; Writing—original draft, F.L.; Writing—review & editing, Q.Z., C.S. and Y.L. All authors have read and agreed to the published version of the manuscript.

Funding: This research received no external funding.

Institutional Review Board Statement: Not applicable.

Informed Consent Statement: Not applicable.

Data Availability Statement: Not applicable.

Conflicts of Interest: The authors declare no conflict of interest.

References

1. Li, W.; Liu, J.; Zeng, J.; Leong, Y.; Elsworth, D.; Tian, J.; Li, L. A fully coupled multidomain and multiphysics model for evaluation of shale gas extraction. *Fuel* **2020**, *278*, 118214. [CrossRef]
2. Curtis, J.B. Fractured Shale-Gas Systems. *AAPG Bull.* **2002**, *86*, 1921.
3. Ma, X.; Xie, J.; Yong, R.; Zhu, Y. Geological characteristics and high production control factors of shale gas reservoirs in Silurian Longmaxi Formation, southern Sichuan Basin, SW China. *Petrol. Explor. Dev.* **2020**, *47*, 901–915. [CrossRef]
4. Hou, P.; Gao, F.; He, J.; Liu, J.; Xue, Y.; Zhang, Z. Shale gas transport mechanisms in inorganic and organic pores based on lattice Boltzmann simulation. *Energy Rep.* **2020**, *6*, 2641–2650. [CrossRef]
5. Yu, H.; Zhu, Y.; Jin, X.; Liu, H.; Wu, H. Multiscale simulations of shale gas transport in micro/nano-porous shale matrix considering pore structure influence. *J. Nat. Gas Sci. Eng.* **2019**, *64*, 28–40. [CrossRef]
6. Zhang, S.; Jia, B.; Zhao, J.; Pu, H. A diffuse layer model for hydrocarbon mass transfer between pores and organic matter for supercritical CO_2 injection and sequestration in shale. *Chem. Eng. J.* **2021**, *406*, 126746. [CrossRef]
7. Ross, D.J.K.; Marc Bustin, R. The importance of shale composition and pore structure upon gas storage potential of shale gas reservoirs. *Mar. Petrol. Geol.* **2009**, *26*, 916–927. [CrossRef]
8. Chen, S.; Zhang, C.; Li, X.; Zhang, Y.; Wang, X. Simulation of methane adsorption in diverse organic pores in shale reservoirs with multi-period geological evolution. *Int. J. Coal Sci. Technol.* **2021**, *8*, 844–855. [CrossRef]
9. Guan, M.; Liu, X.; Jin, Z.; Lai, J. The heterogeneity of pore structure in lacustrine shales: Insights from multifractal analysis using N2 adsorption and mercury intrusion. *Mar. Petrol. Geol.* **2020**, *114*, 104150. [CrossRef]
10. Chandra, D.; Vishal, V. A critical review on pore to continuum scale imaging techniques for enhanced shale gas recovery. *Earth-Sci. Rev.* **2021**, *217*, 103638. [CrossRef]
11. Chalmers, G.R.; Bustin, R.M.; Power, I.M. Characterization of gas shale pore systems by porosimetry, pycnometry, surface area, and field emission scanning electron microscopy/transmission electron microscopy image analyses: Examples from the Barnett, Woodford, Haynesville, Marcellus, and Doig units. *AAPG Bull.* **2012**, *96*, 1099–1119.
12. Li, Y.; Chen, J.; Yang, J.; Liu, J.; Tong, W. Determination of shale macroscale modulus based on microscale measurement: A case study concerning multiscale mechanical characteristics. *Petrol. Sci.* **2022**, *19*, 1262–1275. [CrossRef]
13. Xu, S.; Gou, Q.; Hao, F.; Zhang, B.; Shu, Z.; Lu, Y.; Wang, Y. Shale pore structure characteristics of the high and low productivity wells, Jiaoshiba shale gas field, Sichuan Basin, China: Dominated by lithofacies or preservation condition? *Mar. Petrol. Geol.* **2020**, *114*, 104211. [CrossRef]
14. Wang, Y.; Zhu, Y.; Chen, S.; Li, W. Characteristics of the Nanoscale Pore Structure in Northwestern Hunan Shale Gas Reservoirs Using Field Emission Scanning Electron Microscopy, High-Pressure Mercury Intrusion, and Gas Adsorption. *Energy Fuels* **2014**, *28*, 945–955. [CrossRef]
15. Loucks, R.G.; Reed, R.M.; Ruppel, S.C.; Jarvie, D.M. Morphology, Genesis, and Distribution of Nanometer-Scale Pores in Siliceous Mudstones of the Mississippian Barnett Shale. *J. Sediment. Res.* **2009**, *79*, 848–861. [CrossRef]

16. Zhou, S.; Yan, D.; Tang, J.; Pan, Z. Abrupt change of pore system in lacustrine shales at oil- and gas-maturity during catagenesis. *Int. J. Coal Geol.* **2020**, *228*, 103557. [CrossRef]

17. Chen, S.; Li, X.; Chen, S.; Wang, Y.; Gong, Z.; Zhang, Y. A new application of atomic force microscopy in the characterization of pore structure and pore contribution in shale gas reservoirs. *J. Nat. Gas Sci. Eng.* **2021**, *88*, 103802. [CrossRef]

18. Wang, M.; Yu, Q. Pore structure characterization of Carboniferous shales from the eastern Qaidam Basin, China: Combining helium expansion with low-pressure adsorption and mercury intrusion. *J. Petrol. Sci. Eng.* **2017**, *152*, 91–103. [CrossRef]

19. Tiwari, P.; Deo, M.; Lin, C.L.; Miller, J.D. Characterization of oil shale pore structure before and after pyrolysis by using X-ray micro CT. *Fuel* **2013**, *107*, 547–554. [CrossRef]

20. Li, Y.; Yang, J.; Pan, Z.; Tong, W. Nanoscale pore structure and mechanical property analysis of coal: An insight combining AFM and SEM images. *Fuel* **2020**, *260*, 116352. [CrossRef]

21. Zhang, S.; Li, Y.; Pu, H. Studies of the storage and transport of water and oil in organic-rich shale using vacuum imbibition method. *Fuel* **2020**, *266*, 117096. [CrossRef]

22. Li, Y.; Zhang, C.; Tang, D.; Gan, Q.; Niu, X.; Wang, K.; Shen, R. Coal pore size distributions controlled by the coalification process: An experimental study of coals from the Junggar, Ordos and Qinshui basins in China. *Fuel* **2017**, *206*, 352–363. [CrossRef]

23. Zhang, X.; Liu, C.; Zhu, Y.; Chen, S.; Wang, Y.; Fu, C. The characterization of a marine shale gas reservoir in the lower Silurian Longmaxi Formation of the northeastern Yunnan Province, China. *J. Nat. Gas Sci. Eng.* **2015**, *27*, 321–335. [CrossRef]

24. Zhang, R.; Liu, S.; Wang, Y. Fractal evolution under in situ pressure and sorption conditions for coal and shale. *Sci. Rep.* **2017**, *7*, 8971. [CrossRef]

25. Klaver, J.; Desbois, G.; Urai, J.L.; Littke, R. BIB-SEM study of the pore space morphology in early mature Posidonia Shale from the Hils area, Germany. *Int. J. Coal Geol.* **2012**, *103*, 12–25. [CrossRef]

26. Tong, S.; Dong, Y.; Zhang, Q.; Elsworth, D.; Liu, S. Quantitative Analysis of Nanopore Structural Characteristics of Lower Paleozoic Shale, Chongqing (Southwestern China): Combining FIB-SEM and NMR Cryoporometry. *Energy Fuels* **2017**, *31*, 13317–13328. [CrossRef]

27. Yang, F.; Ning, Z.; Wang, Q.; Liu, H. Pore structure of Cambrian shales from the Sichuan Basin in China and implications to gas storage. *Mar. Petrol. Geol.* **2016**, *70*, 14–26. [CrossRef]

28. Zhao, J.; Jin, Z.; Hu, Q.; Liu, K.; Liu, G.; Gao, B.; Liu, Z.; Zhang, Y.; Wang, R. Geological controls on the accumulation of shale gas: A case study of the early Cambrian shale in the Upper Yangtze area. *Mar. Petrol. Geol.* **2019**, *107*, 423–437. [CrossRef]

29. Liu, X.; Xiong, J.; Liang, L. Investigation of pore structure and fractal characteristics of organic-rich Yanchang formation shale in central China by nitrogen adsorption/desorption analysis. *J. Nat. Gas Sci. Eng.* **2015**, *22*, 62–72. [CrossRef]

30. Iqbal, O.; Padmanabhan, E.; Mandal, A.; Dvorkin, J. Characterization of geochemical properties and factors controlling the pore structure development of shale gas reservoirs. *J. Petrol. Sci. Eng.* **2021**, *206*, 109001. [CrossRef]

31. Liu, B.; Gao, Y.; Liu, K.; Liu, J.; Ostadhassan, M.; Wu, T.; Li, X. Pore structure and adsorption hysteresis of the middle Jurassic Xishanyao shale formation in the Southern Junggar Basin, northwest China. *Energy Explor. Exploit.* **2021**, *39*, 761–778. [CrossRef]

32. Lukens, W.W.; Schmidt-Winkel, P.; Zhao, D.; Feng, J.; Stucky, G.D. Evaluating Pore Sizes in Mesoporous Materials: A Simplified Standard Adsorption Method and a Simplified Broekhoff–de Boer Method. *Langmuir* **1999**, *15*, 5403–5409. [CrossRef]

33. Bertier, P.; Schweinar, K.; Stanjek, H.; Ghanizadeh, A.; Clark, C.R.; Busch, A. On the use and abuse of N2 physisorption for the characterization of the pore structure of shales. In *The Clay Minerals Society Workshop Lectures Series*; Clay Mineral Society: Boulder, CO, USA, 2016; pp. 151–161.

34. Guo, X.; Qin, Z.; Yang, R.; Dong, T.; He, S.; Hao, F.; Yi, J.; Shu, Z.; Bao, H.; Liu, K. Comparison of pore systems of clay-rich and silica-rich gas shales in the lower Silurian Longmaxi formation from the Jiaoshiba area in the eastern Sichuan Basin, China. *Mar. Petrol. Geol.* **2019**, *101*, 265–280. [CrossRef]

35. Wang, F.; Guo, S. Influential factors and model of shale pore evolution: A case study of a continental shale from the Ordos Basin. *Mar. Petrol. Geol.* **2019**, *102*, 271–282. [CrossRef]

36. Sun, W.; Zuo, Y.; Wu, Z.; Liu, H.; Zheng, L.; Wang, H.; Shui, Y.; Lou, Y.; Xi, S.; Li, T.; et al. Pore characteristics and evolution mechanism of shale in a complex tectonic area: Case study of the Lower Cambrian Niutitang Formation in Northern Guizhou, Southwest China. *J. Petrol. Sci. Eng.* **2020**, *193*, 107373. [CrossRef]

37. Zapata, Y.; Sakhaee-Pour, A. Modeling adsorption–desorption hysteresis in shales: Acyclic pore model. *Fuel* **2016**, *181*, 557–565. [CrossRef]

38. Fan, C.; Nguyen, V.; Zeng, Y.; Phadungbut, P.; Horikawa, T.; Do, D.D.; Nicholson, D. Novel approach to the characterization of the pore structure and surface chemistry of porous carbon with Ar, N_2, H_2O and CH_3OH adsorption. *Microporous Mesoporous Mater.* **2015**, *209*, 79–89. [CrossRef]

39. Sun, C.; Nie, H.; Dang, W.; Chen, Q.; Zhang, G.; Li, W.; Lu, Z. Shale Gas Exploration and Development in China: Current Status, Geological Challenges, and Future Directions. *Energy Fuels* **2021**, *35*, 6359–6379. [CrossRef]

40. Chen, X.; Fan, J.; Zhang, Y.; Wang, H.; Chen, Q.; Wang, W.; Liang, F.; Guo, W.; Zhao, Q.; Nie, H.; et al. Subdivision and Delineation of the Wufeng and Lungmachi Black shales in the Subsurface areas of the Yangtze Platform. *J. Stratigr.* **2015**, *39*, 351–358.

41. Liang, F.; Zhang, Q.; Lu, B.; Jiang, W.; Xiong, X.; Peng, C.; Ren, J.; Ling, P.; Chao, M. Lithofacies and Distribution of Wufeng-Longmaxi Organic-rich Shale and Its Impact on Shale Gas Production in Weiyuan Shale Gas Play, Southern Sichuan Basin, China. *Acta Sedimentol. Sin.* **2021**, 1–14. [CrossRef]

42. Zou, C.; Dong, D.; Wang, S.; Li, J. Geological characteristics and resource potential of shale gas in China. *Petrol. Explor. Dev.* **2010**, *37*, 641–653. [CrossRef]

43. Sun, H. COMPASS: An ab Initio Force-Field Optimized for Condensed-Phase ApplicationsOverview with Details on Alkane and Benzene Compounds. *J. Phys. Chem. B* **1998**, *102*, 7338–7364. [CrossRef]

44. Fan, C.; Do, D.D.; Nicholson, D. On the Cavitation and Pore Blocking in Slit-Shaped Ink-Bottle Pores. *Langmuir* **2011**, *27*, 3511–3526. [CrossRef] [PubMed]

45. Neimark, A.V.; Ravikovitch, P.I.; Vishnyakov, A. Adsorption hysteresis in nanopores. *Phys. Rev. E* **2000**, *62 Pt 2A*, R1493–R1496. [CrossRef]

46. Bernard, S.; Wirth, R.; Schreiber, A.; Schulz, H.; Horsfield, B. Formation of nanoporous pyrobitumen residues during maturation of the Barnett Shale (Fort Worth Basin). *Int. J. Coal Geol.* **2012**, *103*, 3–11. [CrossRef]

47. Curtis, M.E.; Cardott, B.J.; Sondergeld, C.H.; Rai, C.S. Development of organic porosity in the Woodford Shale with increasing thermal maturity. *Int. J. Coal Geol.* **2012**, *103*, 26–31. [CrossRef]

48. Milliken, K.L.; Rudnicki, M.; Awwiller, D.N.; Zhang, T. Organic matter-hosted pore system, Marcellus Formation (Devonian), Pennsylvania. *AAPG Bull.* **2013**, *97*, 177–200. [CrossRef]

49. Lash, G.G.; Blood, D. Geochemical and textural evidence for early (shallow) diagenetic growth of stratigraphically confined carbonate concretions, Upper Devonian Rhinestreet black shale, western New York. *Chem. Geol.* **2004**, *206*, 407–424. [CrossRef]

50. Knapp, L.J.; Ardakani, O.H.; Uchida, S.; Nanjo, T.; Otomo, C.; Hattori, T. The influence of rigid matrix minerals on organic porosity and pore size in shale reservoirs: Upper Devonian Duvernay Formation, Alberta, Canada. *Int. J. Coal Geol.* **2020**, *227*, 103525. [CrossRef]

51. Yang, Y.; Aplin, A.C. A permeability-porosity relationship for mudstones. *Mar. Petrol. Geol.* **2010**, *27*, 1692–1697. [CrossRef]

52. Ye, Y.; Luo, C.; Liu, S.; Xiao, C.; Ran, B.; Sun, W.; Yang, D.; Luba, J.; Zeng, X. Characteristics of Black Shale Reservoirs and Controlling Factors of Gas Adsorption in the Lower Cambrian Niutitang Formation in the Southern Yangtze Basin Margin, China. *Energy Fuels* **2017**, *31*, 6876–6894. [CrossRef]

53. Zhao, J.; Jin, Z.; Hu, Q.; Liu, K.; Jin, Z.; Hu, Z.; Nie, H.; Du, W.; Yan, C.; Wang, R. Mineral composition and seal condition implicated in pore structure development of organic-rich Longmaxi shales, Sichuan Basin, China. *Mar. Petrol. Geol.* **2018**, *98*, 507–522. [CrossRef]

54. Chalmers, G.R.L.; Bustin, R.M. Lower Cretaceous gas shales in northeastern British Columbia; Part I, Geological controls on methane sorption capacity. *Bull. Can. Pet. Geol.* **2008**, *56*, 1–21. [CrossRef]

55. Zhu, H.; Ju, Y.; Huang, C.; Qi, Y.; Ju, L.; Yu, K.; Li, W.; Feng, H.; Qiao, P. Petrophysical properties of the major marine shales in the Upper Yangtze Block, south China: A function of structural deformation. *Mar. Petrol. Geol.* **2019**, *110*, 768–786. [CrossRef]

56. Chalmers, G.R.L.; Ross, D.J.K.; Bustin, R.M. Geological controls on matrix permeability of Devonian Gas Shales in the Horn River and Liard basins, northeastern British Columbia, Canada. *Int. J. Coal Geol.* **2012**, *103*, 120–131. [CrossRef]

57. Dong, T.; Harris, N.B.; McMillan, J.M.; Twemlow, C.E.; Nassichuk, B.R.; Bish, D.L. A model for porosity evolution in shale reservoirs: An example from the Upper Devonian Duvernay Formation, Western Canada Sedimentary Basin. *AAPG Bull.* **2019**, *103*, 1017–1044. [CrossRef]

58. Xi, Z.; Tang, S.; Wang, J. The reservoir characterization and shale gas potential of the Niutitang formation: Case study of the SY well in northwest Hunan Province, South China. *J. Petrol. Sci. Eng.* **2018**, *171*, 687–703. [CrossRef]

59. Jia, A.; Hu, D.; He, S.; Guo, X.; Hou, Y.; Wang, T.; Yang, R. Variations of Pore Structure in Organic-Rich Shales with Different Lithofacies from the Jiangdong Block, Fuling Shale Gas Field, SW China: Insights into Gas Storage and Pore Evolution. *Energy Fuels* **2020**, *34*, 12457–12475. [CrossRef]

60. Liang, M.; Wang, Z.; Gao, L.; Li, C.; Li, H. Evolution of pore structure in gas shale related to structural deformation. *Fuel* **2017**, *197*, 310–319. [CrossRef]

61. Ji, L.; Zhang, T.; Milliken, K.L.; Qu, J.; Zhang, X. Experimental investigation of main controls to methane adsorption in clay-rich rocks. *Appl. Geochem.* **2012**, *27*, 2533–2545. [CrossRef]

62. Ju, Y.; Sun, Y.; Tan, J.; Bu, H.; Han, K.; Li, X.; Fang, L. The composition, pore structure characterization and deformation mechanism of coal-bearing shales from tectonically altered coalfields in eastern China. *Fuel* **2018**, *234*, 626–642. [CrossRef]

63. Wang, G. Deformation of organic matter and its effect on pores in mud rocks. *AAPG Bull.* **2020**, *103*, 21–36. [CrossRef]

64. Sun, M.; Zhang, L.; Hu, Q.; Pan, Z.; Yu, B.; Sun, L.; Bai, L.; Fu, H.; Zhang, Y.; Zhang, C.; et al. Multiscale connectivity characterization of marine shales in southern China by fluid intrusion, small-angle neutron scattering (SANS), and FIB-SEM. *Mar. Petrol. Geol.* **2020**, *112*, 104101. [CrossRef]

65. Zhu, H.; Ju, Y.; Huang, C.; Han, K.; Qi, Y.; Shi, M.; Yu, K.; Feng, H.; Li, W.; Ju, L.; et al. Pore structure variations across structural deformation of Silurian Longmaxi Shale: An example from the Chuandong Thrust-Fold Belt. *Fuel* **2019**, *241*, 914–932. [CrossRef]

66. Zhu, H.; Ju, Y.; Qi, Y.; Huang, C.; Zhang, L. Impact of tectonism on pore type and pore structure evolution in organic-rich shale: Implications for gas storage and migration pathways in naturally deformed rocks. *Fuel* **2018**, *228*, 272–289. [CrossRef]

*energies*

Article

The Study on Diagenetic Characteristics of Coal Measures Sandstone Reservoir in Xishanyao Formation, Southern Margin of the Junggar Basin

Aobo Zhang [1,2], Shida Chen [1,2,*], Dazhen Tang [1,2,*], Shuling Tang [1,2], Taiyuan Zhang [1,2], Yifan Pu [1,2] and Bin Sun [3]

1   School of Energy Resources, China University of Geosciences (Beijing), Beijing 100083, China; zhangaobo1994@outlook.com (A.Z.); tangshuling@cugb.edu.cn (S.T.); zhangtaiyuan@cugb.edu.cn (T.Z.); 3006200027@cugb.edu.cn (Y.P.)
2   Coal Reservoir Laboratory of National Engineering Research Center of Coalbed Methane Development & Utilization, Beijing 100083, China
3   Unconventional Research Institute, Research Institute of Petroleum Exploration and Development, Beijing 100083, China; sbin@petrochina.com.cn
*   Correspondence: Shida.Chen@cugb.edu.cn (S.C.); tang@cugb.edu.cn (D.T.)

Abstract: The reservoir physical properties, pore types, diagenetic characteristics and reservoir quality controlling effect of the Xishanyao formation coal measure sandstone in the southern margin of the Junggar basin were discussed in this study based on thin section observation, high pressure mercury injection, low-temperature nitrogen adsorption and scanning electron microscope observation. The result shows that the porosity and permeability of the sandstone are generally low with a medium-high texture maturity and low compositional maturity. The sandstone storage space is mainly composed of residual intergranular pores, secondary dissolution pores, inter-crystalline pores and micro-fractures. The diagenetic stage of coal measure sandstone is in the mesodiagenesis A1-A2 stage, and their diagenetic interaction types mainly include compaction, cementation and dissolution. The reservoir quality of the coal measure sandstone deteriorates by compaction due to high matrix content and plastic debris content. Because of the large amounts of organic acids generated during the thermal evolution of the coal measure source rock, the coal measure sandstone suffers from strong dissolution. The secondary dissolution pores formed by the massive dissolution of feldspar, lithic fragments and early carbonate cementation in the sandstone significantly improved the reservoir quality. In the coal measure sandstone, clay mineral cementation is the most developed cementation form, followed by quartz cementation and carbonate cementation. Although kaolinite cementation and dolomite cementation can generate a small number of inter-crystalline pores, cementation deteriorates the reservoir quality. The Xishanyao formation coal measure sandstone formed in a lacustrine-delta environment, and its composition and texture make it susceptible to the influence of compaction and dissolution during diagenesis.

Keywords: tight sandstone; coal measure; southern margin of Junggar basin; diagenetic process; reservoir quality control; reservoir forming

Citation: Zhang, A.; Chen, S.; Tang, D.; Tang, S.; Zhang, T.; Pu, Y.; Sun, B. The Study on Diagenetic Characteristics of Coal Measures Sandstone Reservoir in Xishanyao Formation, Southern Margin of the Junggar Basin. *Energies* **2022**, *15*, 5499. https://doi.org/10.3390/en15155499

Academic Editor: Rajender Gupta

Received: 29 June 2022
Accepted: 26 July 2022
Published: 29 July 2022

Publisher's Note: MDPI stays neutral with regard to jurisdictional claims in published maps and institutional affiliations.

1. Introduction

The Junggar basin is abundant in coalbed methane (CBM) resources, of which the predicted resource of CBM with a burial depth of less than 2000 m is 3.11×10^{12} m^3, and the southern margin of the Junggar basin (SJB) has the best preservation conditions for CBM [1]. The main coal-bearing strata in the SJB are lower-middle Jurassic, and the sedimentary environment mainly includes alluvial fans, fan deltas, braided river deltas, and shore shallow lakes [1–3]. The SJB has characteristics of thick and multiple coal seams developed with a laterally widely distribution and frequent interbedded sandstone, shale, and coal seams in the vertical direction which is favorable for the co-accumulation of coal measure gases [4–10] (CMG, including CBM, coal measure tight sandstone gas and coal measure shale gas). The precedent of successful development in the Ordos basin [5,11,12] and the maturation of CMG symbiotic accumulation theory [7,8] have made the industry aware of the great potential in CMG development. The reservoir quality is crucial for the efficient development of coal measure tight sandstone gas, and regional differences in diagenetic environment and tectonic events lead to different factors affecting reservoir quality [13–19]. In order to efficiently develop the abundant tight sandstone that is contained in the coal measures in the SJB, it is necessary to conduct research on the petrological characteristics and reservoir quality of coal measures sandstone.

The tight sandstone reservoirs generally undergo a complex diagenesis process, resulting in progressive tightness during burial and thermal evolution [13–18]. The effects of reservoir quality include sedimentary, diagenesis and epigenesis, and the specific influencing factors include clastic composition, depositional environment, sedimentary facies, burial temperature and pressure [19,20]. Sedimentation controls the grain size, sorting, and arrangement of particles, as well as the mineral transformation and sedimentary structure [16,21–23]. The influence of diagenesis on the sandstone is mainly in terms of temperature, pH, pressure, etc., and is presented in the form of compaction, dissolution, cementation, and metasomatism [19,21,24]. Compaction is one of the main factors leading to the reduction of porosity and permeability [25], and the reservoir quality is particularly affected by the degree of compaction and the content of plastic particles (such as lithic fragments, mica, argillaceous debris) and matrix [19,26]. Cementation generally leads to the plugging of pore throats and seriously reduces permeability [27–32], but at the same time, it can also restrain the deterioration of reservoir quality to a certain extent: The inter-crystalline pores formed by the clay minerals cementation can provide storage space for sandstones [31,33,34]; the clay film covering the particle surface resists compaction and inhibits the development of other types of cement [13,35], thereby preserving residual pores; the clay film coating the particle surface can resist compaction and inhibit the development of other types of cement; the quartz cementation and carbonate cementation in the eodiagenesis can improve the compaction resistance ability of sandstone [15,36,37]. By analyzing the sedimentary environment, composition, and diagenesis characteristics of the coal measure sandstone, the reservoir quality control factors of the coal measure sandstone in the Xishanyao formation can be clarified and the formation and evolution process of the coal measure sandstone reservoir can be explained.

The coal seams of the Xishanyao formation on the southern margin of the Junggar basin are mainly in the low-medium coal rank, with frequent interbedding of sandstones, mudstones and coal seams [1]. The diagenesis of coal measure sandstone is strongly influenced by the thermal evolution of coal seams and dark shale [7,8]. Previous studies on sandstone reservoirs have mainly focused on the mechanism of pore throat formation, quantitative characterization of pore throats, and the relationship between oil and gas intrusion and reservoir diagenesis of sandstone reservoirs in different zones or layers within the basin [38–43]. However, studies on the coal-bearing sandstones of the Middle Jurassic Xishanyao formation have also focused on the evolution of the sedimentary environment, the classification of reservoir types, and the classification and evaluation of reservoirs [2,4,44–47]. As the understanding of the diagenetic sequence of the coal measure sandstone reservoirs in the Xishanyao formation, and the influence on reservoir quality is not comprehensive enough, it is urgent to carry out relevant work. This paper combined techniques including casting thin section observation, scanning electron microscope (SEM), XRD experiment, high-pressure mercury intrusion, low-temperature nitrogen adsorption, and Ro test of coal measure shale to conduct research on the following contents: (1) Determine the characteristics of coal measure sandstone in the Xishanyao formation; (2) clarify the diagenetic evolution sequence of coal measure sandstone in the Xishanyao formation; (3) clarify the controlling effect of sedimentation and diagenesis on the reservoir quality of coal measure sandstone. This study is expected to provide a reference for the exploration and development of tight sandstone gas in the Xishanyao formation coal measures.

2. Geology Background

The Junggar basin is located in the northern part of Xinjiang; it is a large-scale Mesozoic-Cenozoic depression basin with an area of about 13.4×10^4 km^2 and an approximately triangular shape in the plane, with a width of 1120 km from east to west and 800 km from north to south [1–3]. Since the Late Paleozoic, the Junggar basin has experienced Hercynian, Indosinian, Yanshan and Himalaya tectonic movements, and the superposition of multi-phase stress fields has led to its structural complexity [2,44,48]. The Junggar basin can be divided into six structural units: the Luliang uplift, the Ulungu depression, the northern Tianshan piedmont thrust belt, the western uplift, the central depression, and the eastern uplift [3,49] (Figure 1A). The SJB is located in the northern Tianshan piedmont thrust belt, which elongates east-west direction with a north-south width of nearly 200 km (Figure 1B). The northern Tianshan piedmont thrust belt is a multi-phase superimposed and inherited structural belt, and is usually divided into five secondary structural units, including the Sikeshu sag, Qigu fault–fold belt, Huomatu anticlinal zone, Huan anticlinal zone, and Fukang fault zone [50].

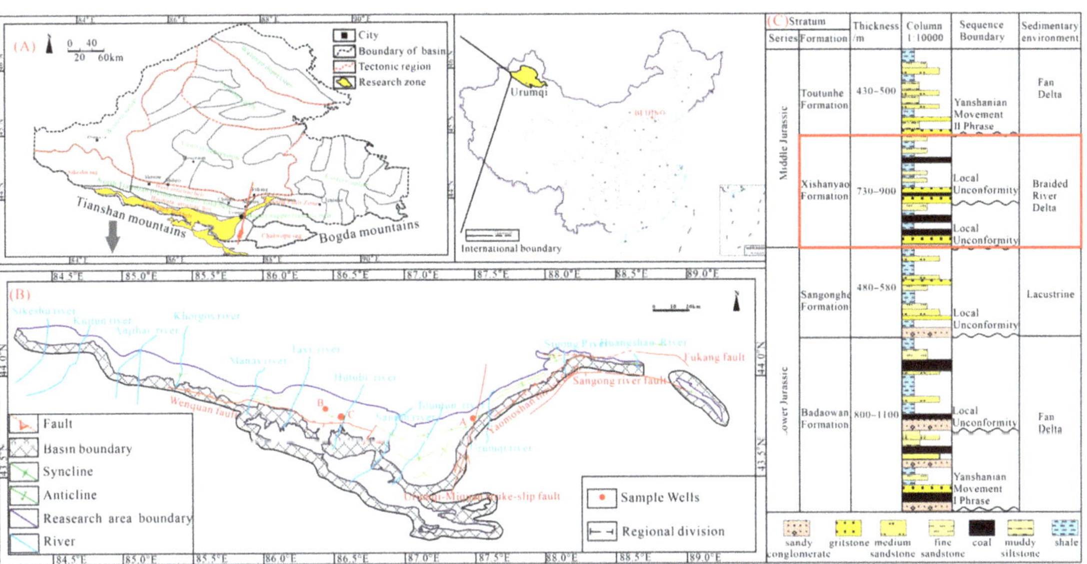

Figure 1. Maps showing: (**A**) Location of the Junggar basin in China (modified by [4]). (**B**) Structural outline map (modified by [10,51]) of the Southern margin of Junggar basin (SJB). (**C**) The stratigraphic column shows the Middle-lower Jurassic Strata in SJB (modified by [51], Xishanyao formation is marked in red).

A major part of the surface of the SJB is covered by quaternary strata. The strata in the SJB area are Permian, Triassic, Jurassic and Palaeocene from old to new (Figure 1C). Among them, the Badaowan formation and Xishanyao formation are the main coal-bearing strata in the study area, which are widely distributed in the SJB (Figure 1C). The coal-bearing strata of the Xishanyao formation were formed in a lacustrine–deltaic sedimentary system [2,3]. The Xishanyao formation strata had multiple coal accumulation centers that were greater than 40 m thick between the present-day Horgos River and Sigong River controlled by tectonic and sedimentary conditions [1].

3. Materials and Methods

Overall, a total of 45 sandstone core samples from three wells in the Xishanyao formation in the SJB were collected and tested (Figure 1B). In addition, 16 shale samples were selected for vitrinite reflectance (Ro) testing to determine diagenetic evolution stages.

3.1. Mineral Composition and Morphology Analysis

A total of 43 casting thin sections were made and placed under a ZEISS Axio Imager II microscope for observation and analysis of the mineral petrological characteristics and pore structure characteristics. What is more, those thin sections have performed image particle size analysis following China's oil and gas industry standard SY/T 5434-2018.

The Scanning Electron Microscope (SEM) observation has been performed on seven sandstone samples by the ZEISS EVO MA 15 scanning electron microscope to clarify their mineral types, pore structure and diagenetic characteristics.

The whole-rock and clay mineral analyses (X-ray diffraction, XRD) were performed on 16 sandstone samples (Table 1) by Rigaku TTR III diffractometer (40 kV, 200 mA and a step size of 8° per minute) equipped with a Cu-Kα radiation source (λ = 1.541841 Å) following the Chinese oil and gas industry standard SY/T 5163-2010.

3.2. Petrophysical Parameters Analysis

The porosity and permeability tests were conducted on 21 sandstone samples (Table 2) using a CoreLab CAT-113 He-porosimeter and a CoreLab CAT-112 gas permeameter, respectively, according to Chinese oil and gas industry standard SY/T5336-2006.

3.3. Pore Structure Characterization Tests

The high-pressure mercury intrusion experiments (HPMI) were carried out on 27 samples by a Micromeritics Autopore IV 9505 instrument to characterize the complexity and heterogeneity of the pore throat structure (Table 2) following the China Petroleum and Natural Gas Industry Standard SY/T 5346-2012. The instrument automatically records the pressure, pore size, intrusion volume and surface area with the increase in the mercury injected amount.

The low-pressure N_2 gas adsorption (N_2GA) experiments were conducted on 23 crushed sandstone samples (Table 2) using V-Sorb 2800P series specific surface area and pore size analyzers (temperature 77 K, pressure < 127 kPa). The specific surface area and total pore volume were calculated using the BET model [52] and the BJH model [53] according to the N_2GA data, respectively.

Table 1. The result of XRD experiments of sandstones in the Xishanyao formation.

Well	Strata	Depth (m)	Whole Rock (%)									Clay Mineral (%)					
			Quartz	K-Felspar	Plagioclase	Calcite	Dolomite	Apatite	Ankerite	Siderite	Clay Mineral	S	I/S	Illite	Kaolinite	Chlorite	C/S
A	J_2x	303	54.3	5	25.5			1.2			14		62	36	2		
A	J_2x	601	50	0.6	16.6						32.8		62	15	23		
A	J_2x	653.5	43.6	0.2	21			1.1			34.1		50	16	34		
A	J_2x	669	43.3	0.5	10					15.9	30.3		61	15	24		
A	J_2x	673	38.4	0.7	16.4			4.5			40		48	26	26		
B	J_2x	919	48.6	3.1	14			8.5			25.8		31	39	23	7	
B	J_2x	922.5	47.9	9.9	19.1			6.7			16.4		23	32	40	5	
B	J_2x	935	41.1	10.8	22.9			7.2			18		25	32	43		
B	J_2x	994.5	61.5	1.2	29.1	2					6.2		38	17	35	10	
B	J_2x	1055	48.2	7.6	29.7						14.5		43	30	27		
B	J_2x	1090	41.5	8.3	15.3		10.4		12		12.5		20	30	50		
C	J_2x	967	44.9	12.2	25.3			2.6			15		16	34	50		
C	J_2x	987	41.7	7.8	15.7		10	4.8			20		15	32	47	6	
C	J_2x	988	43.4	6.2	17.4		23.8	1.6			7.6		17	39	44		
C	J_2x	1011	50.1	6.4	16			2.8			24.7		24	38	38		
C	J_2x	1061	58.7	11.6	17.7			0.5			11.5		25	32	43		

Table 2. The physical properties and pore structure characterization of coal measure sandstones in the Xishanyao Formation.

Well	Depth	Sorting Effi-cient	Matrix Con-tent (%)	Mean Grain Size (mm)	Permeability (mD)	Porosity (%)	Pore Throat Radius (μm)		Maximum Injected Mercury Saturation (%)	Mercury Withdrawal Efficiency (%)	Pore Volume ($\times 10^{-5}$ mL/g)			Specific Surface Area (m^2/g)
							Mean Value	Median Value			Micropores	Mesopores	Macropores	
A	303	1.211	4.5	0.203	0.53900	7.481	0.319	0.241	95.00	55.86	52	269	138	1.242
A	305	1.264	4.5	0.225							36	308	99	1.062
A	305.61	1.232	4.2	0.241										
A	308	1.806	4.2	0.274	0.15270	10.86					43	252	128	0.982
A	355	1.931	10.5	0.035			0.616	0.331	95.96	51.51				
A	380	1.381	8.5	0.048										
A	404	1.273	4	0.153										
A	407.05	1.056	5	0.116										
A	507.4	0.933	5	0.098										
A	573	0.973	5	0.114										
A	601				0.00039	3.476	0.012	0.010	95.05	47.92	214	1348	313	6.099
A	653.5	1.248	10	0.071	0.00068	4.058	0.015	0.012	91.93	40.28	167	967	227	4.905
A	655	0.707	9	0.041										
A	669						0.012	0.008	91.27	49.21	175	1050	295	5.124
A	670	1.252	20	0.028										
A	673	1.246	20	0.033			0.015	0.012	94.79	42.01	244	1091	325	6.53
B	841	1.258	4	0.301										
B	889.9	0.723	4	0.222										
B	919	1.283	8	0.087	0.00154	1.889	0.037	0.022	85.18	55.95	43	295	57	1.15
B	921	0.838	3.8	0.118	0.00600	1.400	0.050	0.029	89.89	53.23				
B	922	1.239	10	0.081										
B	922.5	1.411	15	0.054			0.013	0.006	75.05	73.00	43	265	62	0.93
B	934.1	0.816	4	0.161										
B	935	1.164	8	0.080	0.00016	1.30	0.015	0.006	65.74	57.21	32	134	69	0.64
B	937.5	1.079	8	0.131			0.088	0.015	71.34	65.06	27	159	52	0.51
B	968.7	1.282	4.5	0.148										
B	987	1.348	5	0.131	0.00710	1.93	0.081	0.044	88.09	56.03	30	147	37	0.54
B	1018.5	0.996	4.5	0.181										
B	1034	1.080	3.5	0.274	0.04400	3.91	0.162	0.080	93.50	34.31				
B	1055	0.992	4	0.386	0.28900	3.11	0.413	0.124	91.23	51.48	44	239	161	0.77
B	1090	0.896	4.5	0.136	0.00320	1.68	0.010		34.91	25.29	23	176	56	0.41
B	1091	0.897	15	0.062	0.00280	1.38					60	366	103	1.50

Table 2. *Cont.*

Well	Depth	Sorting Effi-cient	Matrix Con-tent (%)	Mean Grain Size (mm)	Permeability (mD)	Porosity (%)	Pore Throat Radius (μm) Mean Value	Median Value	Maximum Injected Mercury Saturation (%)	Mercury Withdrawal Efficiency (%)	Pore Volume ($\times 10^{-5}$ mL/g) Micropores	Mesopores	Macropores	Specific Surface Area (m^2/g)
C	967	0.757	4	0.167	0.16100	5.03	0.233	0.134	94.28	48.85	59	220	75	1.33
C	970.2	0.784	3.5	0.199	0.09800	4.65	0.195	0.155	94.18	39.73				
C	984	0.800	6.5	0.146	0.06200	2.37	0.091	0.050	92.75	34.68	41	262	175	0.92
C	984	1.384	5	0.168			0.229	0.075	95.72	53.83				
C	987	0.855	4.6	0.158	0.02430	3.75	0.112	0.050	92.35	47.85	46	258	109	1.05
C	988	0.875	4.5	0.168			0.076	0.023	91.57	50.46	36	145	46	0.64
C	988.5	1.471	7.8	0.135	0.01700	2.32	0.070	0.039	96.19	36.86				
C	991	1.111	15	0.026	3.42200	1.51								
C	994	0.900	5.2	0.039	0.00300	1.25	0.011	0.006	77.98	39.81				
C	1011	0.961	5	0.091	0.00469	4.87	0.042	0.015	91.39	73.21	43	289	170	0.79
C	1046	0.917	4	0.240	0.11680	2.73	0.134	0.039	93.33	53.09	53	365	94	1.49
C	1047	0.935	4.2	0.223			0.209	0.064	91.84	40.24	61	282	112	1.36
C	1061	0.873	4	0.415	0.20200	3.26	0.334	0.096	89.98	52.25	26	152	60	0.78

4. Results

4.1. Composition and Texture of Sandstone

The sandstone of the Xishanyao formation in the SJB is mainly feldspar lithic sandstone [47], and its compositional maturity is relatively low. The provenance of coal measure sandstone reservoirs comes from the Tianshan Mountains, and the parent rock types are middle-low-grade metamorphic rocks and middle-acid magmatic rocks [46,54].

According to the results of the image particle size analysis experiment. The particle size of sandstone is concentrated between silt and medium grains (mean grain size is 0.026–0.415 mm, Figure 2C), with a poor to medium sorting degree (sorting coefficient is 0.707–1.931), sub-angular to sub-round particle roundness, and high matrix content (3.5–20%, avg on 6.8%, Table 2). This evidence indicates that the texture maturity of coal measure sandstone is medium-high mature.

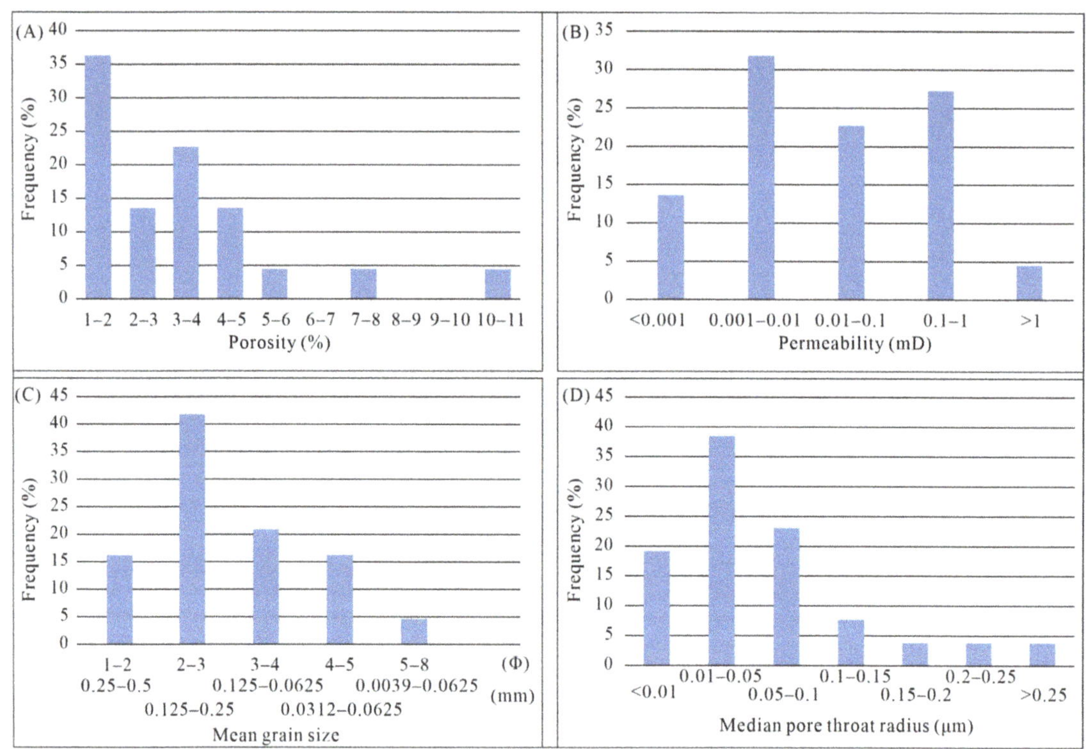

Figure 2. Frequency distribution histogram of porosity (**A**), permeability (**B**), average grain size (**C**) and median grain size radius (**D**).

The proportion of detrital particles in coal measure sandstone was analyzed through an XRD experiment: the content of quartz in the detrital grains was the highest, ranging from 38.4% to 61.5% (avg 47.3%); feldspar minerals included plagioclase and K-feldspar. Plagioclase content is slightly higher than K-feldspar, which is in the range between 10 and 29.7% (avg 19.5%); K-feldspar content is between 0.2 and 12.2% (avg 5.8%); apatite can be detected in many samples, and its content is between 0.5 and 8.5% (avg 3.5%); some samples can detect dolomite (shows in three samples, accounting for 10–23.8%), ankerite (shows in one sample, accounting for 12%) and siderite (shows in one sample, accounting for 15.9%); clay mineral content is between 2.0 and 29.8% (avg 7.8%).

The XRD test results of clay minerals show that the relative contents of clay minerals in the Xishanyao formation coal measure sandstone (Figure 3, Table 1): the content of illite is between 15 and 39% (avg 28.9%); the content of illite/smectite (I/S) mixed layer is between 15 and 62% (avg 35%); the kaolinite content is between 2 and 50% (avg 34%); chlorite can be found in three samples, with its content between 5 and 10% (avg 7%). With the increase in burial depth, the content of kaolinite and illite gradually increases, while the content of I/S decreases (Figure 3). The mixed layer ratio(S%) of I/S decreased from 30% to 10%, indicating the I/S is gradually transformed from disordered to ordered mixed layers.

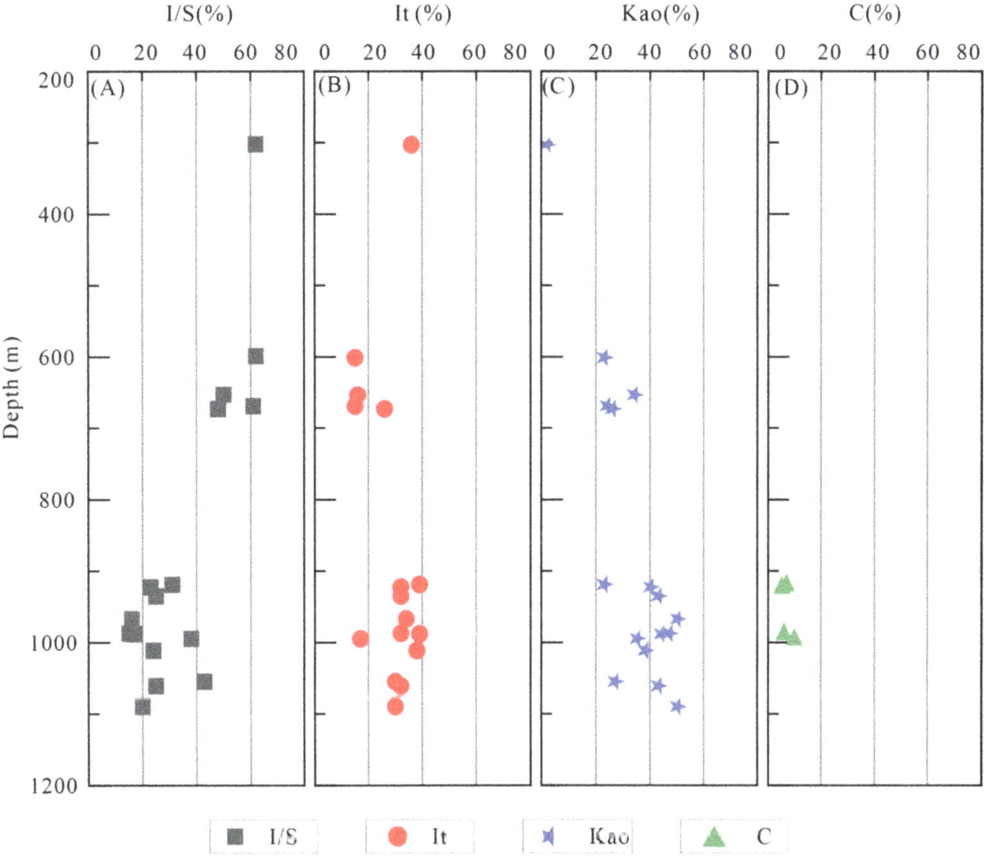

Figure 3. Relative contents of different clay minerals with increasing depth. (**A**) illite/smectite mixed layers; (**B**) illite; (**C**) Kaolinite; (**D**) chlorite.

4.2. Physical Properties of Coal Measure Sandstone

The core porosity of the coal measure sandstone ranges from 1.25 to 10.86% (avg of 3.37%, Figure 2A), and the permeability of the core samples is in the range of 0.00016–0.539 mD (avg on 0.083 mD, Figure 2B) which is mainly composed of lots of low permeability and low porosity sandstone and several tight sandstones. As the depth increases, the porosity and permeability of the sandstone samples decrease accordingly, and the correlation between porosity and depth is more obvious.

4.3. Characterization of Sandstone Storage Space

4.3.1. Pore Structure Characterization

The results of the HPMI test can provide effective parameters for evaluating pore connectivity [15,36,37]. The average pore-throat radius of the coal-bearing sandstones of the Xishanyao formation ranges from 0.010 to 0.616 um, with a mean value of 0.133 um (Table 2); the discharge pressure ranges from 0.26 to 13.79 Mpa; the median pore-throat radius ranges from 0.006 to 0.331 um (Figure 2D); the maximum injected mercury saturation ranges from 34.91% to 96.19%, and the mercury withdrawal efficiency ranges from 25.29% to 73.21 (Table 2). Overall, the sandstone pore-throat radius and the connectivity in this area are moderate. Combined with the analysis of the physical property results, the relationship between the permeability and the median pore-throat radius is better than the relationship between the permeability with mean grain size or porosity (Figure 4A–C), which also indicates that the pore-throat radius is an important factor in determining the seepage capacity [55].

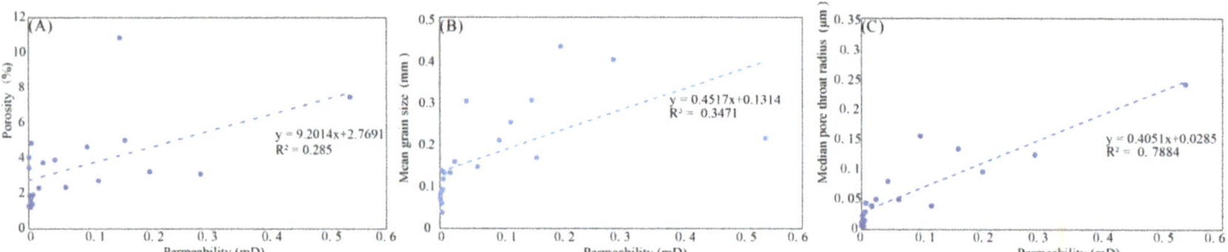

Figure 4. Plots of the relationship between permeability with porosity (**A**), average grain size (**B**), and median pore throat radius (**C**).

Nitrogen adsorption experiments show that the mesopore volume of the Xishanyao formation coal measure sandstone is higher than others, and the adsorption volume is 134×10^{-5}–1348×10^{-5} mL/g (avg on 393×10^{-5} mL/g) followed by the macropore with an adsorption volume of 37×10^{-5}–325×10^{-5} mL/g (avg on 128×10^{-5} mL/g). The micropore adsorption volume is the smallest, which is 23×10^{-5}–244×10^{-5} mL/g (avg on 69×10^{-5} mL/g). The specific surface area is between 0.409 and 6.53 m^2/g, with an average value of 1.77 m^2/g. The origin of pores in different diameters is quite different, which means pore volume distribution can help analyze the influence of diagenesis [56].

4.3.2. Types of Storage Space

The observations from casting thin sections and SEM of the coal measure sandstone of the Xishanyao formation show that the storage space in the study area mainly includes primary pores, secondary pores and fractures. The proportion of primary pores is relatively small, mainly residual primary pores (RIPs), which are triangular or polygonal shapes (Figure 5C,D). The secondary pores are relatively abundant, mainly composed of intergranular dissolved pores (such as carbonate cemented dissolution pores, Figure 5B,C), intragranular dissolved pores (such as feldspar and lithic fragment dissolution pores) (Figure 5A) and a small amount of inter-crystalline pores (kaolinite inter-crystalline pores, dolomite inter-crystalline pores, Figure 5E,F). Among them, the surface of the intergranular dissolved pores is relatively irregular, while the intragranular dissolved pores are mostly located on the surface of the debris particles. The fractures in the sandstone include intragranular fractures and intergranular fractures, which are mostly formed under strong compaction. The existence of these fractures improves reservoir permeability to a certain extent.

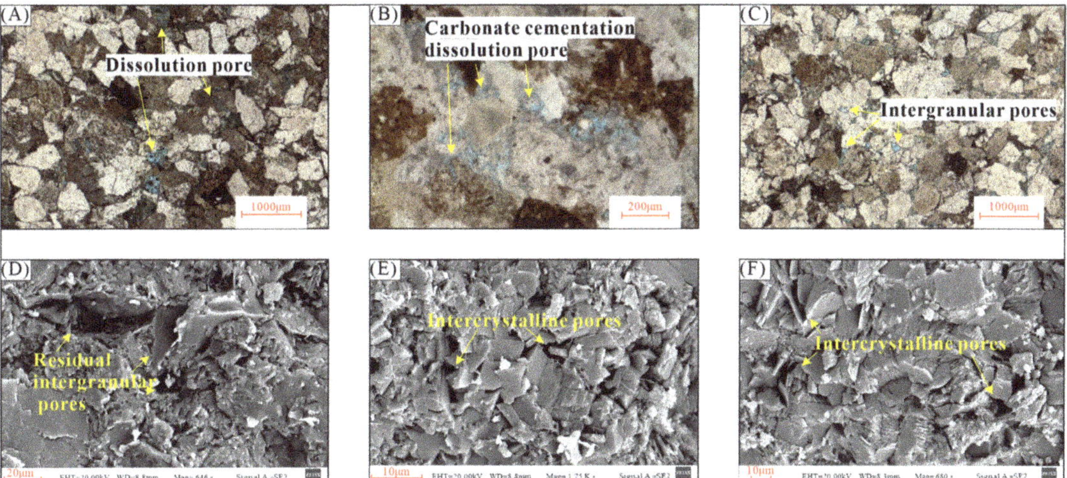

Figure 5. Pore types of coal measure sandstone. (**A**) Intergranular dissolution pores; (**B**) Cementation dissolution pores; (**C**) Intergranular pores; (**D**) Residual intergranular pores; (**E**) Dolomite inter-crystalline pores; (**F**) Kaolinite inter-crystalline pores.

4.4. Diagenesis Feature

4.4.1. Compaction

The constantly increasing weight of the overlying water and sediments results in compaction degree increases, and in the sediments appear a series of phenomena, such as interlayer water discharge, volume reduction, porosity reduction, and permeability decline [29,37,57,58]. Mechanical compaction causes the rearrangement and breaking of the rigid grains and deforms plastic particles, which will reduce the pore space in the reservoir and inhibits cementation and dissolution [25,59]. Further intensification of the pressure leads to the occurrence of chemical compaction, which is characterized by suture contact. The compaction phenomenon observed in coal measure sandstones indicates intense compaction during the diagenetic stage: particle contact types include linear contact (Figure 6A), concave–convex contact (Figure 6A,D) and suture contact (Figure 6F); rigid particles, such as quartz and feldspar are broken to produce cracks (Figure 6A,F–H); feldspar particles are in a semi-oriented arrangement (Figure 6D,F); plastic components, such as mica, plastic lithic fragments, and carbonaceous fragments are bent and deformed (Figure 6D–F).

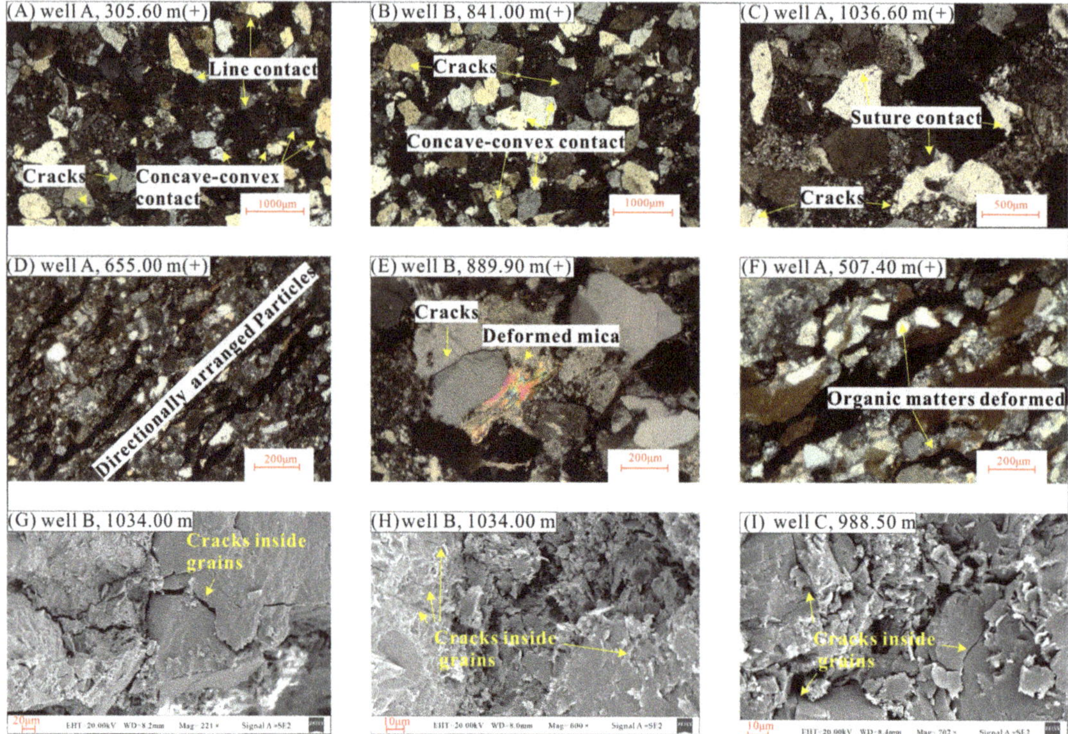

Figure 6. Characteristics of compaction show on casting thin sections and SEM images in the southern margin of Junggar basin. (**A**,**B**) the line contact and the concave–convex contact of the detrital particles, and intragranular fissures can be found on the detrital particles; (**C**) the concave–convex contact and suture contact of detrital particles; (**D**) the compacted plastic particles and semi-directional distributed rigid particle; (**E**) the deformed mica; (**F**) organic debris and boulder clay filled pores and fissures; (**G–I**) the intragranular fractures. In the thin section picture, (+) means orthogonally polarized light.

4.4.2. Cementation

- Quartz cement;

Quartz cement commonly occurs in acidic diagenetic environments, usually in the form of overgrowths around detrital quartz grains [13,60], and the fluid inclusions, irregular erosion and remnant clay films often occur in the boundaries of quartz grains and quartz cementation (Figure 5A,C). The quartz grains in the study area are rarely observed to be encrusted with clay films, which makes the overgrowth hard to identify. In eodiagenesis, quartz cementation tends to be associated with I/S mineral transformations, and in the mesodiagenesis stage, the quartz cementation may result from the chemical compaction of quartz grains and the dissolution of feldspar and lithic fragments [5,19,23,61–64].

- Clay mineral cement;

The clay minerals' cementation is the major cementation type in the SJB which usually occurs in the form of authigenic shapes filling the pores or covering the surface of the grains [13,65] (Figure 7D). The authigenic shape of kaolinite is usually booklet and helminthoid filled in intergranular pores and dissolution pores [5] (Figure 7E,G), and its origin is usually associated with the dissolution of feldspar and aluminosilicate lithic fragments [43]. The illite in the SJB is generally in a flaky, fibrous shape and covers the surface of the grains [55] (Figure 7E–I). In addition, illite was also found on the surface of dolomite crystals, which showed the persistence of illite cementation in diagenesis (Figure 7H). The I/S is the product of the smectite/illite transformation, usually in the form of honeycomb-filled pores [65] (Figure 7F).

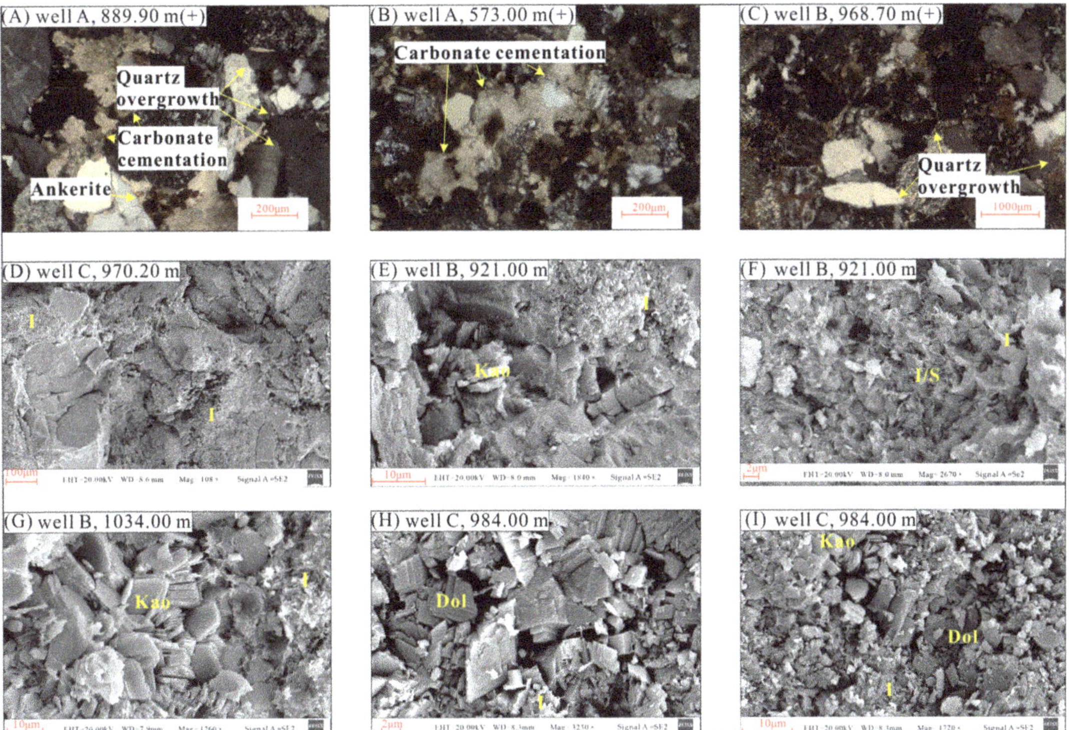

Figure 7. The characteristics of cementation are shown in casting thin sections and SEM images in the study area. (**A**) the cementation of micrite carbonate and dolomite, and the quartz overgrowth, (**B**) micrite carbonate cemented filled intergranular pores; (**C**) quartz overgrowth, clay minerals cement filling pores, and sericitization of volcanic lithic fragments; (**D**) illite cement in close contact with detrital particles; (**E**) kaolinite and illite cement filled pores; (**F**) illite and I/S cement fill intergranular pores; (**G**) the kaolinite inter-crystalline pores; (**H**) dolomite fills intergranular pores and forms inter crystalline pores, while the illite is attached to the surface dolomite; (**I**) kaolinite and illite and dolomite cement fill the intergranular pores. In the thin section picture, (+) means orthogonally polarized light.

- Carbonate cement;

During the mesodiagenesis A2 stage, the organic acid content was greatly reduced due to high-temperature decomposition and consumption of dissolution, resulting in a gradual weakening of the acidic diagenetic environment. The CO_3^{2-} generated from the dissolution of early period carbonate cement and the thermal evolution of organic matter combines with the Ca^{2+}, Fe^{2+} and Mg^{2+} ions that come from the dissolution of early period carbonate cement to form the dolomite and ankerite cement [65]. These late period carbonate cement (dolomite and ankerite) filled the residual intergranular pores (Figure 7A,B,H,I).

4.4.3. Dissolution

The coal measure sandstones of the Xishanyao formation have been strongly modified by dissolution due to the large amounts of organic acids released during the thermal evolution of the coal measure source rock (mainly from coal seams and some from dark shales). The dissolution occurs mainly in feldspars, Equations (1)–(3), lithic fragments (Figure 8A,B,E,F), and carbonate cement (Figure 8C,D). Compared with the dissolution of feldspar, the dissolution of carbonate cement is often accompanied by a lower chemical equilibrium constant, thus requiring a lower PH value of the diagenetic environment [41,66]. There are numerous particle dissolutions pores (Figure 8A,B,G–I), casting pores (Figure 8A,B,E,F) and intergranular pores produced by cement dissolution (Figure 8C,D) in the casting thin sections and SEM image.

$$2KAlSi_3O_8 + 2H^+ + H_2O = Al_2Si_2O_5(OH)_4 + 4SiO_2 + 2K^+, \tag{1}$$

$$2NaAlSi_3O_8 + 2H^+ + H_2O = Al_2Si_2O_5(OH)_4 + 4SiO_2 + 2Na^+, \tag{2}$$

$$CaAl_2Si_2O_8 + 2H^+ + H_2O = Al_2Si_2O_5(OH)_4 + Ca^{2+} + 3Al^{3+}, \tag{3}$$

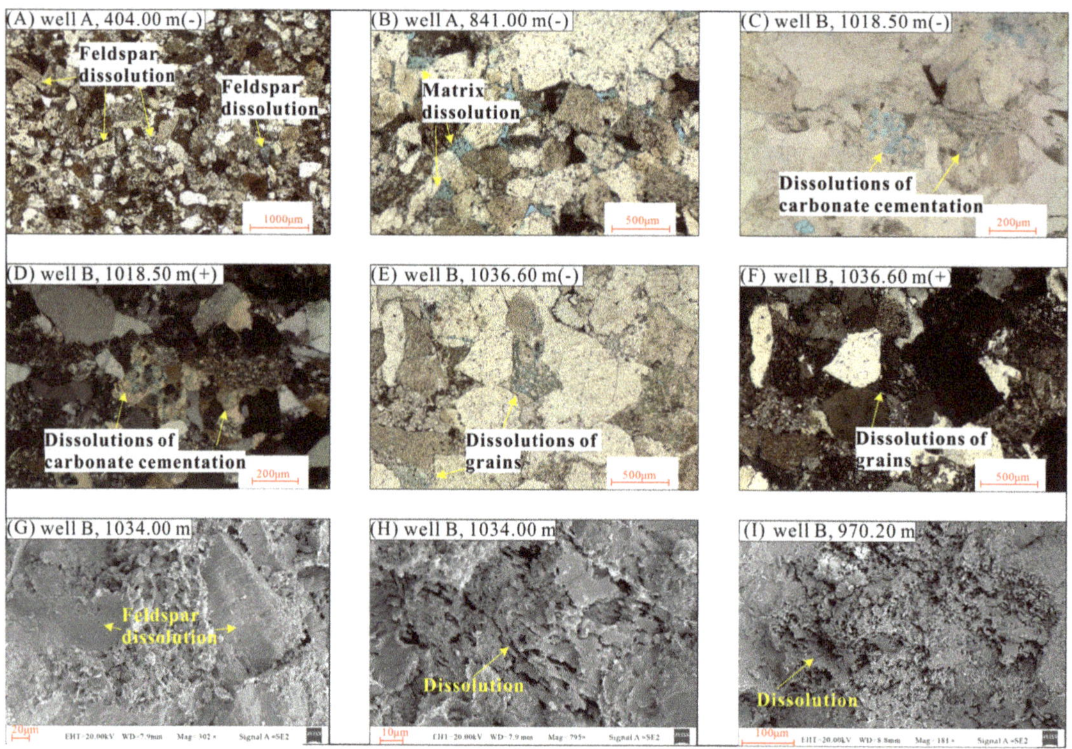

Figure 8. Characteristics of dissolution show on casting thin sections and SEM images in the study area. (**A**) dissolution of feldspar and volcanic lithic fragments. (**B**) dissolution of feldspar, lithic fragments and matrix; (**C,D**) the dissolution pores that are formed by dissolution of micrite carbonate cementation; (**E,F**) feldspar dissolves along the cleavage plane, forming dissolution pores and fractures; (**G**) dissolution on feldspar particle surface with the kaolinite filling the intergranular pores; (**H,I**) directional dissolution of feldspar particle. In the thin section picture, (+) means orthogonally polarized light and (-) means plane polarized light.

5. Discussion

5.1. Paragenetic Sequence of Coal Measure Sandstone Diagenesis

5.1.1. Determination of Diagenetic Stage

The clarification of diagenetic stages should be based on paleo-temperature (T), vitrinite reflectance (Ro), maximum pyrolysis peak temperature (Tmax), and smectite mixed layer ratio (S%) of I/S, spatial and temporal assemblages of authigenic minerals and the contact relationships between particles [5,22,67]. The Ro of coal measure shale in the Xishanyao formation in the study area ranges from 0.63–0.89%. The clay minerals of the coal measure sandstone are dominated by I/S, kaolinite and illite, with the mixed layer ratio (S%) of I/S ranging from 10–30% (avg on 18%). The compaction strongly influences the coal measure sandstone, with contact relationships composed of line contact, concave–convex contact, and suture contact. The reconstructed thermal history of the Jurassic strata, which is conducted by Wang et al. [68], found that the maximum burial depth of the Xishanyao formation exceeded 3000 m and the paleotemperature exceeded 100 °C. According to the diagenetic stage classification standard of clastic rocks (China's oil and gas industry standard SY/T5477-2003), it can be comprehensively determined that the diagenetic stage of the Xishanyao formation coal measure sandstone is in the mesodiagenesis A1 and A2 stage (Figure 8).

5.1.2. Diagenetic Sequences of Coal Measure Sandstone

Combining with the thermal and burial history of the SJB, referring to the division of diagenetic stages in clastic rocks of China's oil and gas industry standard SY/T5477-2003, the diagenetic evolution of the Xishanyao formation coal measure sandstone in the area is summarized as follows:

- Syndiagenesis

In this stage, the diagenetic environment has not been completely disconnected from the overlying water. The organic matter deposited in the peat swamp generates CO_2 under the action of microorganisms and dissolves in water, making the diagenetic environment weakly acidic. Due to the acidic nature of the sedimentary water, the surface of the detrital particles is rarely coated by a chlorite clay film. Due to the strong hydrodynamic environment during the sedimentary period, the coal measure sandstone contains a certain amount of plastic micrite carbonate and carbonaceous fragments will be blended into the sandstone.

- Eodiagenesis (at the depth < 2 km, temperature < 70 °C, and Ro < 0.5%)

The type and degree of diagenesis in the eodiagenesis stage will affect the type of diagenesis and the physical properties of the reservoir in the later diagenetic stage [22]. At this stage, the diagenetic environment is weakly acidic, and the diagenesis types include mechanical compaction, cementation (mainly argillaceous cementation) and dissolution, and the diagenetic environment gradually changes to a closed system. The compaction significantly reduces primary porosity [69,70], due to the shallow burial depth and the high subsidence rate. The rock is in a weakly consolidated and semi-consolidated state, and the contact relationships of the clastic particles change from point contact to point-line contact. Pore types of the coal measure sandstone in this stage are mainly primary pores, and a small amount of micrite carbonate cement fills the pores.

With the increase in temperature, the coal seams and dark shale gradually released organic acid and CO_2 [64], which makes the diagenetic environment gradually change to a weak acid. The feldspar and detrital grains began to dissolve slightly and form dissolution pores (Figure 9).

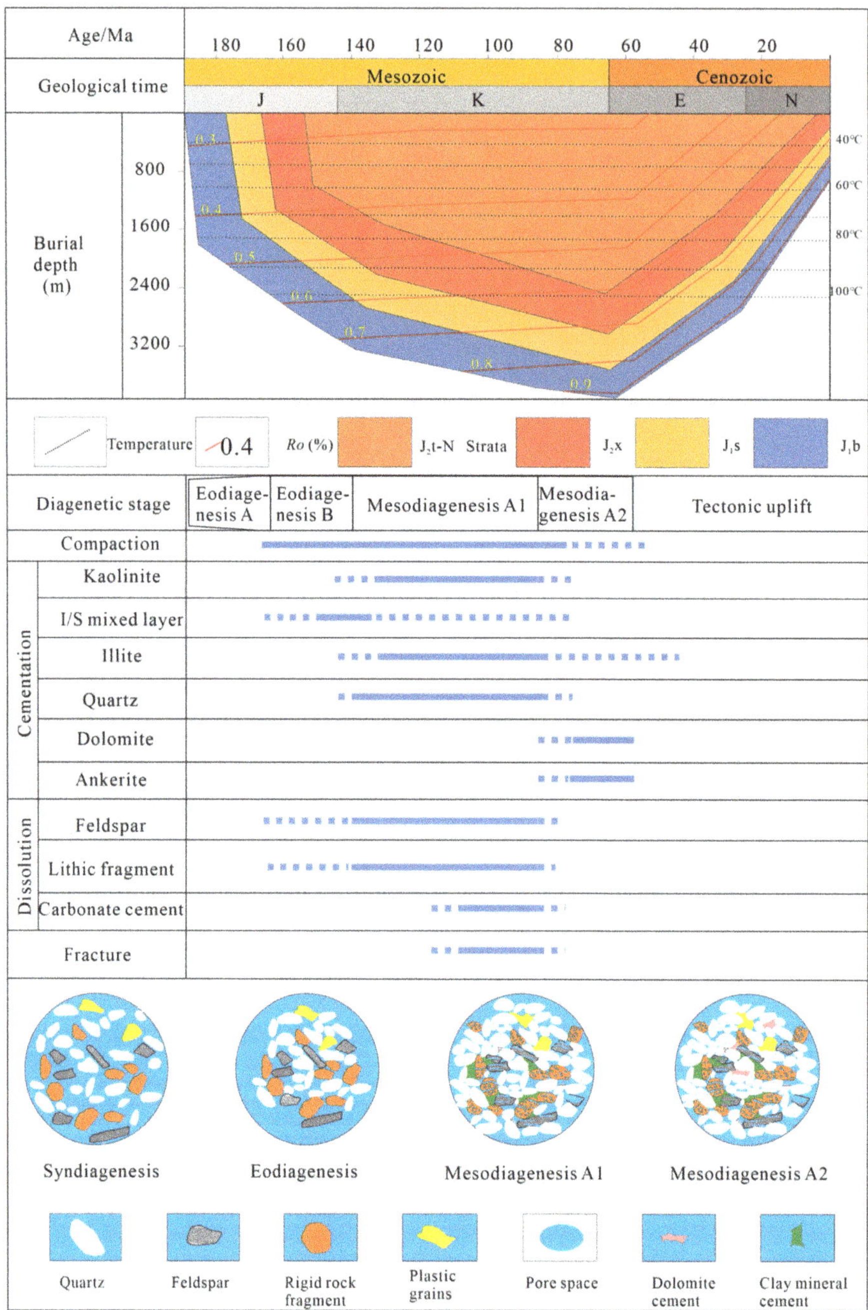

Figure 9. The paragenetic sequence and pore evolution of the Xishanyao formation in the SJB. The thermal evolution history is based on the study of Wang et al. [68] with some modifications. Solid blue lines represent probable timing based on observed diagenetic and mineralization phases. While the dashed blue lines represent inferred or not well-constrained diagenetic and mineralization phases. J_2x = Xishanyao formation; I/S = illite/smectite; Ro = vitrinite reflectance.

- Mesodiagenesis (at the depth > 2 km, temperature > 70 °C, and Ro > 0.5%)

The effect of compaction in the mesodiagenesis A stage was extremely strong, and the compaction type gradually shifted from mechanical compaction to chemical compaction. The contact relationships change from point contact and line contact to line contact, concave–convex contact and suture contact during this stage, along with a large proportion of brittle minerals broken up. The dissolution pores formed in the eodiagenesis stage are damaged under the compaction, and the damage of compaction to the quality of the reservoir is further strengthened. The mesodiagenesis A stage can be further divided into two substages, A1 and A2, according to temperature and Ro. These two substages have certain differences in diagenetic environment and diagenetic reaction:

In the mesodiagenesis A1 stage, the temperature was 70–90 °C, and the Ro was 0.5–0.7%, in an acidic diagenetic environment. The amount of organic acid generated in coal measures reaches the highest level at this temperature [71], the diagenetic environment is acidic, and the tensity of dissolution reached the highest level (Figure 9). The massive dissolution of feldspar, aluminosilicate lithic fragments and early micrite carbonate cement formed a large number of dissolution pores and mold pores. Because of the relatively closed diagenetic system, the dissolution products cannot be discharged. This has led to the kaolinite filling pores and quartz overgrowth (part of the SiO_2 comes from the chemical compaction).

In the Mesodiagenesis A2 stage, the temperature was 90–130 °C, and the Ro was 0.7–1.3%. The acidic diagenetic environment is gradually weakened due to the decomposition of organic acids. The effect of dissolution decreases gradually, while the cementation is gradually increases. The quartz overgrowth gradually decreases, which is manifested as second level quartz cementation. Late period carbonate cementation (dolomite and ankerite) gradually filled the pore space (Figure 9). The clay minerals appear abundantly and fill the pores, further reducing the porosity.

5.2. Diagenetic Control of Coal Measure Rvoir Quality

Reservoir quality is influenced by diagenetic activities, such as compaction, cementation, dissolution, recrystallization, and metasomatism [20,72]. For the Xishanyao formation coal measure sandstone, compaction, dissolution and cementation are the main diagenetic controlling factors for reservoir quality.

5.2.1. The Influence of Compaction on Reservoir Quality

The compaction is one of the main factors that cause the reduction of intergranular pore volume [73] and the densification of coal measure sandstone reservoirs. Plastic particles in the sandstone have a weak compaction resistance, while the rigid particles have a high compaction resistance ability [61,74]. The coal measure sandstones of the Xishanyao formation contain a high content of plastic grains, such as volcanic lithic fragments, muddy gravel and carbonaceous fragments, which makes its compaction resistance ability weak.

The coal measure sandstone reservoirs are strongly affected by mechanical compaction, which is characteristic of the close contacted clastic particles, the ruptured rigid particles, and the deformed plastic particles. Chemical compaction commonly happens in SJB due to the high intensity of compaction, which may lead to particle suture contact and reduced interparticle space. The intense mechanical and chemical compaction has a great negative effect on the physical properties of the sandstone reservoir.

5.2.2. The Influence of Cementation on Reservoir Quality

The coal measure sandstone of the Xishanyao formation in the SJB is mainly clay mineral cement, followed by quartz cement and carbonate cement. The development of these cementations is one of the main factors leading to the significant decline in reservoir quality [74–77].

The closed diagenetic system of coal measure sandstone prevents the dissolution products from being discharged in time, which also leads to the development of authigenic kaolinite cementation and quartz cementation. The developed kaolinite cement fills the pore space in large quantities, but its loose inter-crystalline structure contains inter-crystalline pores that can resist compaction [78]. In addition, illite and I/S mainly fill intergranular pores and pore throats, thereby reducing the reservoir quality of the sandstone [65,79,80]. In summary, different types of clay minerals occur widely and fill or divide the pore space, resulting in tortuous pore throats, reducing pore connectivity, and weakening seepage capacity [13,37,65,79,80].

The content of volcanic lithic fragments in coal measure sandstone is relatively high, which may inhibit the quartz overgrowth to some extent [77]. It can be observed that the overgrowth level of the coal measure sandstone is shown in the second stage (Figure 5A,C). The quartz overgrowth in the eodiagenesis stage can help the sandstone better resist compaction. In the mesodiagenesis, pressure solution, feldspar dissolution and clay mineral transformation produced many SiO_2 (aq), which developed into the quartz cement [13,81,82].

The carbonate cement is locally developed in coal measure sandstones. In the eodiagenesis stage, several micrite carbonate particles blended in the coal measure sandstone, which blocked and filled the intergranular pores. In the mesodiagenesis A2 stage, the late period dolomite and ankerite cements appeared in the sandstone, which blocked the intergranular pores of the sandstone and reduced the reservoir quality (Figure 5A,H,I).

5.2.3. The Influence of Dissolution on Reservoir Quality

The dissolution in diagenesis has a positive effect on reservoir physical properties [13]. The coal seam and dark shale begin to produce organic acids through the decomposition of plant residues during the eodiagenesis stage [71] and reaching a peak in the mesodiagenesis stage (temperature range from 80 °C to 140 °C, [37,83]). The amount of organic acid generated in the coal measure is much higher than in other source rocks [5,7,84].

The sedimentary environment makes the sandstone have good primary physical properties and pore connectivity, and the weak dissolution in the eodiagenesis stage provides a seepage channel for the migration of acidic fluids in the mesodiagenesis stage. The large proportion of aluminosilicate minerals, such as feldspar and volcanic in coal measure sandstone provides a good material basis for the dissolution [5]. Dissolution is most intense in mesodiagenesis. During this period, the thermal evolution of organic matter releases a large number of organic acids into the diagenetic system of the sandstone, causing the soluble minerals to selectively dissolve and form a large number of intragranular dissolved pores and intergranular dissolution pores [35]. Overall, the diagenetic stage of the coal measure sandstone in this area is mainly in the A1 and A2 stages of the mesodiagenesis, which happens to be the most intense stage of organic acid dissolution. The dissolution has brought a great positive effect on the physical properties of the reservoir.

5.2.4. The Origin of Authigenic Clay Minerals in Coal Measure Sandstone

Clay mineral cement has a great influence on the physical properties of coal measure sandstone reservoirs. The coal measure sandstone has a high content of kaolinite (2–50%, avg 34%) and some chlorite (5–10% in 4 samples) which is abnormal in coal measure. The discussion of their origins could offer more information to help us understand the diagenetic characteristic of coal measure sandstone.

The origin of kaolinite in authigenic minerals mainly includes (1) weathering of medium-acid magmatic rocks and metamorphic rocks during sedimentation [85]; (2) dissolution of volcanic lithic fragments and feldspar during diagenesis, Equations (1)–(3) [13]; (3) the alternation of unstable illite into stable kaolinite in an acidic environment [5]. Based on the evidence of relatively strong dissolution in the sandstone, the high content of volcanic lithic fragment and feldspar and the developed authigenic quartz in sandstone, it can be considered that the kaolinite in the study area mainly comes from the dissolution.

The formation of chlorite requires Fe^{2+} and Mg^{2+}. Based on the source of Fe^{2+} and Mg^{2+}, the formation of chlorite mainly includes (1) direct precipitation in the delta sedimentary environment due to the difference in electrolytes between rivers and lakes during sedimentation [86,87]; (2) formed by the transformation of smectite under an alkaline environment, and this process is marked by the existence of chlorite/smectite mixed layers [87]; (3) the interaction of feldspar and Fe^{2+} and Mg^{2+} generated by the dissolution of volcanic lithic fragments Equation (4) [86–88]. However, no chlorite clay film was found during the thin section observation, and no chlorite/smectite mixed layer was detected in the whole-rock and clay mineral analyses result (Table 1), which means the chlorite in the study area did not originate from the first two reasons. Considering the strong dissolution in the study area while the sandstone containing chlorite is in the mesodiagenesis A2 stage, it can be interpreted that the chlorite in the study area is mainly formed by the interaction of feldspar and Fe^{2+} and Mg^{2+} formed by the dissolution of volcanic lithic fragments Equation (4).

$$2KAlSi_3O_8 + 0.4Fe^{2+} + 0.3Mg^{2+} + 1.4H_2O = 0.3(Fe_{14}Mg_{12}Al_{2.5})(Al_{0.7}Si_{3.3})O_{10}(OH)_8 + 2SiO_2 + 0.4H^+ + K^+ \quad (4)$$

(K-feldspar)   (chlorite)

5.3. The Influence of Sedimentation on the Reservoir Quality of Coal Measure Sandstone

The influence of the sedimentation on reservoir quality cannot be well evaluated only by sedimentary facies, which need to be analyzed from the aspects of composition, texture, and diagenetic alteration [19,72,89–91].

The parameters, such as the average particle size, sorting, roundness, and matrix content of the sandstone determine the initial intergranular volume [15,21,32,92,93], and the initial porosity of sediment can reach 40% [94]. Permeability is a function of the matrix and grain size [39,95]. For sandstones in the same diagenetic stage, the larger the average grain size, the better the sorting and roundness; the lower the matrix content, the stronger the compaction resistance, and the better the physical properties of the reservoir will be [96,97]. Reservoirs with good physical properties are conducive to the exchange of fluids, thereby enhancing the strength of cementation and dissolution during diagenesis [67,98].

The coal measures sandstone in the Xishanyao formation is formed in a lacustrine-braided river delta environment [1,3]. The sandstone particles are poor to medium sorted, sub-angular to sub-round roundness, have high matrix content, and have a medium-high texture maturity. Coal measure sandstones with medium-high texture maturity are prone to grain rearrangement (rotation and slippage) during compaction, thereby enhancing the effect of compaction [99].

Differences in the composition of particles lead to differences in physical and chemical diagenesis reactions during diagenesis [67,98,100]. When the quartz content is low, the reservoir quality increases with the increase in the quartz content. If the content of detrital quartz is higher than 75%, the lower content of feldspar will lead to the reduction of dissolved pores, and at the same time, the quartz cementation in the reservoir will be more developed, thereby reducing the porosity and permeability of the reservoir [101,102].

The compositional maturity of coal measure sandstone in the SJB is relatively low, and its provenance is composed of pyroclastic rocks, intermediate-acid magmatic rocks, and metamorphic rocks derived from the Tianshan Mountains [46,54]. Because the provenance is closer to the depositional area, the proportion of feldspar and volcanic lithic fragments in the coal measure sandstone is relatively large. Due to the weak compaction resistance of feldspar, lithic fragments and plastic particles (such as muddy gravel and carbonaceous fragments) [103,104], the coal measure sandstone reservoirs are greatly affected by mechanical compaction. In addition, the abundant organic matter and the high content of feldspar and lithic fragments in the coal measure strata also lead to strong dissolution during the diagenetic stage.

Due to the warm and humid environment during the deposition period of the Xishanyao formation, the coal measures are generally developed. The abundant organic matter generated CO_2 under the action of microbial fermentation and made the water weakly acidic [9]. The weakly acidic water medium condition makes it difficult to coat clay minerals on the grain surface and form chlorite clay films. The absence of the clay film on the grain surface makes the reservoir easily affected by compaction and makes the observation of quartz overgrowth more difficult. In addition to the dissolution during the diagenesis, a part of kaolinite is also formed during the depositional period, and their origins were mainly because of the mineral transformation, illite and chlorite under acidic conditions, and feldspar alteration [5,65,85].

6. Conclusions

The petrological and mineralogical results of the coal measure sandstone samples of the Xishanyao formation in the southern margin of the Junggar basin show that:

1. Coal measure sandstone reservoirs are characterized by a high content of quartz, feldspar and clay minerals, poor to moderate sorting, sub-angular to sub-round roundness, and generally high matrix content. The porosity and permeability of the sandstone layers are low, and multiple samples can be categorized as tight sandstone.

2. The coal measure sandstone is in the mesodiagenesis A1-A2 stage. The effect of compaction in the eodiagenesis was significant, accompanied by a small amount of feldspar and lithic fragments dissolution, kaolinite and illite precipitation, I/S transformation, and quartz cementation. In the mesodiagenesis stage, the influence of compaction is relatively reduced while the I/S transformation level is increased. In the mesodiagenesis A1 stage, the dissolution strength reached the highest level, and the clay mineral and quartz cement were also intense. In the mesodiagenesis A2 stage, the dissolution strength gradually weakened when the late period carbonate cement appeared.

3. The compaction, cementation and dissolution are the three diagenetic factors that control the reservoir quality of the coal measure sandstone. Among them, compaction is the main factor that makes the reservoir quality decline. Dissolution produces massive dissolved pores, which improves the quality of the reservoir significantly. The kaolinite and quartz cementation formed by the precipitation of undischarged dissolution products block the pores and pore throats and deteriorate the reservoir quality.

4. The sedimentary environment influences the texture, composition and diagenetic alteration of coal measure sandstone. The medium-good texture maturity and a large content of volcanic lithic fragments and feldspar of the coal measure sandstone make it susceptible to the influence of compaction and dissolution.

Author Contributions: Conceptualization, D.T. and S.C.; methodology, S.T. and S.C.; software, A.Z., Y.P. and T.Z.; validation, A.Z., Y.P. and D.T.; formal analysis, A.Z.; investigation, A.Z., Y.P., B.S. and T.Z.; resources, D.T.; data curation, A.Z.; writing—original draft preparation, A.Z.; writing—review and editing, A.Z., S.C. and S.T.; visualization, A.Z. and S.T.; supervision, D.T. and S.T.; project administration, D.T. and S.C.; funding acquisition, D.T. and B.S. All authors have read and agreed to the published version of the manuscript.

Funding: The research was financially supported by the National Natural Science Foundation of China (Grant No. 42102205), and the PetroChina Company Limited "14th Five Year Plan" Science and Technology Major Project (Grant No. 2021DJ2306).

Data Availability Statement: Not applicable.

Acknowledgments: We would like to thank the Xinjiang Coalfield Geology Bureau and the No. 156 Coalfield Geological Exploration Team of the Xinjiang Coalfield Geology Bureau for providing access to convenient field work and core samples.

Conflicts of Interest: The authors declare no conflict of interest.

References

1. Tang, D.; Yang, S.; Tang, S.; Tao, S.; Chen, S.; Zhang, A.; Pu, Y.; Zhang, T. Advance on Exploration—Development and Geological Research of Coalbed Methane in the Junggar Basin. *J. China Coal Soc.* **2021**, *46*, 2412–2425. [CrossRef]
2. Li, Y.; Shao, L.; Hou, H.; Tang, Y.; Yuan, Y.; Zhang, J.; Shang, X.; Lu, J. Sequence Stratigraphy, Palaeogeography, and Coal Accumulation of the Fluvio-Lacustrine Middle Jurassic Xishanyao Formation in Central Segment of Southern Junggar Basin, NW China. *Int. J. Coal Geol.* **2018**, *192*, 14–38. [CrossRef]

3. Fu, H.; Tang, D.; Xu, H.; Xu, T.; Chen, B.; Hu, P.; Yin, Z.; Wu, P.; He, G. Geological Characteristics and CBM Exploration Potential Evaluation: A Case Study in the Middle of the Southern Junggar Basin, NW China. *J. Nat. Gas. Sci. Eng.* **2016**, *30*, 557–570. [CrossRef]

4. Fu, H.; Tang, D.; Pan, Z.; Yan, D.; Yang, S.; Zhuang, X.; Li, G.; Chen, X.; Wang, G. A Study of Hydrogeology and Its Effect on Coalbed Methane Enrichment in the Southern Junggar Basin, China. *Bulletin* **2019**, *103*, 189–213. [CrossRef]

5. Li, Y.; Gao, X.; Meng, S.; Wu, P.; Niu, X.; Qiao, P.; Elsworth, D. Diagenetic Sequences of Continuously Deposited Tight Sandstones in Various Environments: A Case Study from Upper Paleozoic Sandstones in the Linxing Area, Eastern Ordos Basin, China. *Bulletin* **2019**, *103*, 2757–2783. [CrossRef]

6. Li, Y.; Yang, J.; Pan, Z.; Meng, S.; Wang, K.; Niu, X. Unconventional Natural Gas Accumulations in Stacked Deposits: A Discussion of Upper Paleozoic Coal-Bearing Strata in the East Margin of the Ordos Basin, China. *Acta Geol. Sin. Engl. Ed.* **2019**, *93*, 111–129. [CrossRef]

7. Qin, Y. Research Progress of Symbiotic Accumulation of Coal Measure Gas in China. *Nat. Gas Ind. B* **2018**, *5*, 466–474. [CrossRef]

8. Zou, C.; Yang, Z.; Huang, S.; Ma, F.; Sun, Q.; Li, F.; Pan, S.; Tian, W. Resource Types, Formation, Distribution and Prospects of Coal-Measure Gas. *Pet. Explor. Dev.* **2019**, *46*, 451–462. [CrossRef]

9. Fu, H.; Li, Y.; Su, X.; Yan, D.; Yang, S.; Wang, G.; Wang, X.; Zhao, W. Environmental Conditions and Mechanisms Restricting Microbial Methanogenesis in the Miquan Region of the Southern Junggar Basin, NW China. *GSA Bull.* **2022**. [CrossRef]

10. Li, Y.; Fu, H.; Yan, D.; Su, X.; Wang, X.; Zhao, W.; Wang, H.; Wang, G. Effects of Simulated Surface Freshwater Environment on in Situ Microorganisms and Their Methanogenesis after Tectonic Uplift of a Deep Coal Seam. *Int. J. Coal Geol.* **2022**, *257*, 104014. [CrossRef]

11. Zhang, L.; Bai, G.; Luo, X.; Ma, X.; Chen, M.; Wu, M.; Yang, W. Diagenetic History of Tight Sandstones and Gas Entrapment in the Yulin Gas Field in the Central Area of the Ordos Basin, China. *Mar. Pet. Geol.* **2009**, *26*, 974–989. [CrossRef]

12. Zou, C.; Zhang, G.; Yang, Z.; Tao, S.; Hou, L.; Zhu, R.; Yuan, X.; Ran, Q.; Li, D.; Wang, Z. Concepts, Characteristics, Potential and Technology of Unconventional Hydrocarbons: On Unconventional Petroleum Geology. *Pet. Explor. Dev.* **2013**, *40*, 413–428. [CrossRef]

13. Yuan, G.; Cao, Y.; Gluyas, J.; Li, X.; Xi, K.; Wang, Y.; Jia, Z.; Sun, P.; Oxtoby, N.H. Feldspar Dissolution, Authigenic Clays, and Quartz Cements in Open and Closed Sandstone Geochemical Systems during Diagenesis: Typical Examples from Two Sags in Bohai Bay Basin, East China. *AAPG Bull.* **2015**, *99*, 2121–2154. [CrossRef]

14. Dutton, S.P.; Loucks, R.G. Reprint of: Diagenetic Controls on Evolution of Porosity and Permeability in Lower Tertiary Wilcox Sandstones from Shallow to Ultradeep (200–6700 m) Burial, Gulf of Mexico Basin, U.S.A. *Mar. Pet. Geol.* **2010**, *27*, 1775–1787. [CrossRef]

15. Lai, J.; Wang, G.; Cai, C.; Fan, Z.; Wang, S.; Chen, J.; Luo, G. Diagenesis and Reservoir Quality in Tight Gas Sandstones: The Fourth Member of the Upper Triassic Xujiahe Formation, Central Sichuan Basin, Southwest China. *Geol. J.* **2018**, *53*, 629–646. [CrossRef]

16. Karim, A.; Pe-Piper, G.; Piper, D.J.W. Controls on Diagenesis of Lower Cretaceous Reservoir Sandstones in the Western Sable Subbasin, Offshore Nova Scotia. *Sediment. Geol.* **2010**, *224*, 65–83. [CrossRef]

17. Wolela, A. Diagenetic Evolution and Reservoir Potential of the Barremian–Cenomanian Debre Libanose Sandstone, Blue Nile (Abay) Basin, Ethiopia. *Cretac. Res.* **2012**, *36*, 83–95. [CrossRef]

18. Zhang, Y.; Pe-Piper, G.; Piper, D.J.W. How Sandstone Porosity and Permeability Vary with Diagenetic Minerals in the Scotian Basin, Offshore Eastern Canada: Implications for Reservoir Quality. *Mar. Pet. Geol.* **2015**, *63*, 28–45. [CrossRef]

19. Taylor, T.R.; Giles, M.R.; Hathon, L.A.; Diggs, T.N.; Braunsdorf, N.R.; Birbiglia, G.V.; Kittridge, M.G.; Macaulay, C.I.; Espejo, I.S. Sandstone Diagenesis and Reservoir Quality Prediction: Models, Myths, and Reality. *AAPG Bull.* **2010**, *94*, 1093–1132. [CrossRef]

20. Bjørlykke, K.; Jahren, J. Open or Closed Geochemical Systems during Diagenesis in Sedimentary Basins: Constraints on Mass Transfer during Diagenesis and the Prediction of Porosity in Sandstone and Carbonate Reservoirs. *AAPG Bull.* **2012**, *96*, 2193–2214. [CrossRef]

21. Morad, S.; Ketzer, J.M.; De Ros, L.F. Spatial and Temporal Distribution of Diagenetic Alterations in Siliciclastic Rocks: Implications for Mass Transfer in Sedimentary Basins. *Sedimentology* **2000**, *47*, 95–120. [CrossRef]

22. El-ghali, M.A.K.; Mansurbeg, H.; Morad, S.; Al-Aasm, I.; Ajdanlisky, G. Distribution of Diagenetic Alterations in Fluvial and Paralic Deposits within Sequence Stratigraphic Framework: Evidence from the Petrohan Terrigenous Group and the Svidol Formation, Lower Triassic, NW Bulgaria. *Sediment. Geol.* **2006**, *190*, 299–321. [CrossRef]

23. Saïag, J.; Brigaud, B.; Portier, É.; Desaubliaux, G.; Bucherie, A.; Miska, S.; Pagel, M. Sedimentological Control on the Diagenesis and Reservoir Quality of Tidal Sandstones of the Upper Cape Hay Formation (Permian, Bonaparte Basin, Australia). *Mar. Pet. Geol.* **2016**, *77*, 597–624. [CrossRef]

24. Burley, S.D.; Kantorowicz, J.D.; Waugh, B. Clastic Diagenesis. *Geol. Soc. Lond. Spec. Publ.* **1985**, *18*, 189–226. [CrossRef]

25. Morad, S.; Al-Ramadan, K.; Ketzer, J.M.; De Ros, L.F. The Impact of Diagenesis on the Heterogeneity of Sandstone Reservoirs: A Review of the Role of Depositional Facies and Sequence Stratigraphy. *AAPG Bull.* **2010**, *94*, 1267–1309. [CrossRef]

26. Yang, S.; Wang, Y.; Zhang, S.; Wang, Y.; Zhang, Y.; Zhao, Y. Controls on Reservoirs Quality of the Upper Jurassic Mengyin Formation Sandstones in Dongying Depression, Bohai Bay Basin, Eastern China. *Energies* **2020**, *13*, 646. [CrossRef]

27. Dashti, R.; Rahimpour-Bonab, H.; Zeinali, M. Fracture and Mechanical Stratigraphy in Naturally Fractured Carbonate Reservoirs-A Case Study from Zagros Region. *Mar. Pet. Geol.* **2018**, *97*, 466–479. [CrossRef]

28. Desbois, G.; Urai, J.L.; Kukla, P.A.; Konstanty, J.; Baerle, C. High-Resolution 3D Fabric and Porosity Model in a Tight Gas Sandstone Reservoir: A New Approach to Investigate Microstructures from Mm- to Nm-Scale Combining Argon Beam Cross-Sectioning and SEM Imaging. *J. Pet. Sci. Eng.* **2011**, *78*, 243–257. [CrossRef]

29. Fisher, Q.J.; Haneef, J.; Grattoni, C.A.; Allshorn, S.; Lorinczi, P. Permeability of Fault Rocks in Siliciclastic Reservoirs: Recent Advances. *Mar. Pet. Geol.* **2018**, *91*, 29–42. [CrossRef]

30. Sakhaee-Pour, A.; Bryant, S.L. Effect of Pore Structure on the Producibility of Tight-Gas Sandstones. *AAPG Bull.* **2014**, *98*, 663–694. [CrossRef]

31. Xiao, L.; Zou, C.; Mao, Z.; Jin, Y.; Shi, Y.; Guo, H.; Li, G. An Empirical Approach of Evaluating Tight Sandstone Reservoir Pore Structure in the Absence of NMR Logs. *J. Pet. Sci. Eng.* **2016**, *137*, 227–239. [CrossRef]

32. Yue, D.; Wu, S.; Xu, Z.; Xiong, L.; Chen, D.; Ji, Y.; Zhou, Y. Reservoir Quality, Natural Fractures, and Gas Productivity of Upper Triassic Xujiahe Tight Gas Sandstones in Western Sichuan Basin, China. *Mar. Pet. Geol.* **2018**, *89*, 370–386. [CrossRef]

33. Stroker, T.M.; Harris, N.B.; Elliott, W.C.; Marion Wampler, J. Diagenesis of a Tight Gas Sand Reservoir: Upper Cretaceous Mesaverde Group, Piceance Basin, Colorado. *Mar. Pet. Geol.* **2013**, *40*, 48–68. [CrossRef]

34. Walderhaug, O.; Eliassen, A.; Aase, N.E. Prediction of Permeability in Quartz-Rich Sandstones: Examples from the Norwegian Continental Shelf and the Fontainebleau Sandstone. *J. Sediment. Res.* **2012**, *82*, 899–912. [CrossRef]

35. Li, P.; Zheng, M.; Bi, H.; Wu, S.; Wang, X. Pore Throat Structure and Fractal Characteristics of Tight Oil Sandstone: A Case Study in the Ordos Basin, China. *J. Pet. Sci. Eng.* **2017**, *149*, 665–674. [CrossRef]

36. Jiu, B.; Huang, W.; Li, Y. An Approach for Quantitative Analysis of Cementation in Sandstone Based on Cathodoluminescence and MATLAB Algorithms. *J. Pet. Sci. Eng.* **2020**, *186*, 106724. [CrossRef]

37. Zhang, Y.; Tian, J.; Zhang, X.; Li, J.; Liang, Q.; Zheng, X. Diagenesis Evolution and Pore Types in Tight Sandstone of Shanxi Formation Reservoir in Hangjinqi Area, Ordos Basin, Northern China. *Energies* **2022**, *15*, 470. [CrossRef]

38. Cao, J.; Hu, W.; Wang, X.; Zhu, D.; Tang, Y.; Xiang, B.; Wu, M. Diagenesis and Elemental Geochemistry under Varying Reservoir Oil Saturation in the Junggar Basin of NW China: Implication for Differentiating Hydrocarbon-Bearing Horizons. *Geofluids* **2015**, *15*, 410–420. [CrossRef]

39. Gao, C.; Ding, X.; Zha, M.; Qu, J. Diagenesis of Unconformity and the Influence on Reservoir Physical Properties: A Case Study of the Lower Jurassic in Xiazijie Area, Junggar Basin, NW China. *J. Petrol. Explor. Prod. Technol.* **2017**, *7*, 659–666. [CrossRef]

40. Tang, Z.; Parnell, J.; Longstaffe, F.J. Diagenesis and Reservoir Potential of Permian–Triassic Fluvial/Lacustrine Sandstones in the Southern Junggar Basin, Northwestern China1. *AAPG Bull.* **1997**, *81*, 1843–1865. [CrossRef]

41. Wang, J.; Zhou, L.; Liu, J.; Zhang, X.; Zhang, F.; Zhang, B. Acid-Base Alternation Diagenesis and Its Influence on Shale Reservoirs in the Permian Lucaogou Formation, Jimusar Sag, Junggar Basin, NW China. *Pet. Explor. Dev.* **2020**, *47*, 962–976. [CrossRef]

42. Wen, H.; Jiang, Y.; Huang, H.; Liu, Y.; Wang, T.; Jiang, Y.; Jin, J.; Qi, L. Provenance and Diagenesis of the Upper Jurassic Qigu Formation Sandstones in the Southern Junggar Basin, NW China. *Arab. J. Geosci.* **2016**, *9*, 452. [CrossRef]

43. Zhou, T.; Wu, C.; Guan, X.; Wang, J.; Jiao, Y. Coupling Relationship between Reservoir Densification and Hydrocarbon Emplacement in the Jurassic Badaowan Formation Sandstone Reservoirs of the Southern Junggar Basin, China. *Geol. J.* **2022**, *57*, 2473–2496. [CrossRef]

44. Bian, W.; Hornung, J.; Liu, Z.; Wang, P.; Hinderer, M. Sedimentary and Palaeoenvironmental Evolution of the Junggar Basin, Xinjiang, Northwest China. *Palaeobio. Palaeoenv.* **2010**, *90*, 175–186. [CrossRef]

45. Hou, H.; Shao, L.; Tang, Y.; Zhao, S.; Yuan, Y.; Li, Y.; Mu, G.; Zhou, Y.; Liang, G.; Zhang, J. Quantitative Characterization of Low-Rank Coal Reservoirs in the Southern Junggar Basin, NW China: Implications for Pore Structure Evolution around the First Coalification Jump. *Mar. Pet. Geol.* **2020**, *113*, 104165. [CrossRef]

46. Fang, Y.; Wu, C.; Guo, Z.; Hou, K.; Dong, L.; Wang, L.; Li, L. Provenance of the Southern Junggar Basin in the Jurassic: Evidence from Detrital Zircon Geochronology and Depositional Environments. *Sediment. Geol.* **2015**, *315*, 47–63. [CrossRef]

47. Guo, H.; Si, X.; Tang, X.; Xu, Y.; Peng, B. Diagenetic Characteristics and Evaluation of Lower-Middle Jurassic Reservoirs in Qigu Fault-Fold Belt, Junggar Basin. *Xinjiang Pet. Geol.* **2020**, *41*, 46–54.

48. Chen, B.; Arakawa, Y. Elemental and Nd-Sr Isotopic Geochemistry of Granitoids from the West Junggar Foldbelt (NW China), with Implications for Phanerozoic Continental Growth. *Geochim. Cosmochim. Acta* **2005**, *69*, 1307–1320. [CrossRef]

49. Zhao, R.; Zhang, J.; Zhou, C.; Zhang, Z.; Chen, S.; Stockli, D.F.; Olariu, C.; Steel, R.; Wang, H. Tectonic Evolution of Tianshan-Bogda-Kelameili Mountains, Clastic Wedge Basin Infill and Chronostratigraphic Divisions in the Source-to-Sink Systems of Permian-Jurassic, Southern Junggar Basin. *Mar. Pet. Geol.* **2020**, *114*, 104200. [CrossRef]

50. Zhou, T.; Wu, C.; Guan, X.; Wang, J.; Zhu, W.; Yuan, B. Effect of Diagenetic Evolution and Hydrocarbon Charging on the Reservoir-Forming Process of the Jurassic Tight Sandstone in the Southern Junggar Basin, NW China. *Energies* **2021**, *14*, 7832. [CrossRef]

51. Tang, S.; Tang, D.; Tao, S.; Sun, B.; Zhang, A.; Zhang, T.; Pu, Y.; Zhi, Y. CO_2-Enriched CBM Accumulation Mechanism for Low-Rank Coal in the Southern Junggar Basin, China. *Int. J. Coal Geol.* **2022**, *253*, 103955. [CrossRef]

52. Brunauer, S.; Emmett, P.H.; Teller, E. Adsorption of Gases in Multimolecular Layers. *J. Am. Chem. Soc.* **1938**, *60*, 309–319. [CrossRef]

53. Barrett, E.P.; Joyner, L.G.; Halenda, P.P. The Determination of Pore Volume and Area Distributions in Porous Substances. I. Computations from Nitrogen Isotherms. *J. Am. Chem. Soc.* **1951**, *73*, 373–380. [CrossRef]

54. Zhou, T.; Wu, C.; Yuan, B.; Shi, Z.; Wang, J.; Zhu, W.; Zhou, Y.; Jiang, X.; Zhao, J.; Wang, J. New Insights into Multiple Provenances Evolution of the Jurassic from Heavy Minerals Characteristics in Southern Junggar Basin, NW China. *Pet. Explor. Dev.* **2019**, *46*, 67–81. [CrossRef]

55. Xi, K.; Cao, Y.; Haile, B.G.; Zhu, R.; Jahren, J.; Bjørlykke, K.; Zhang, X.; Hellevang, H. How Does the Pore-Throat Size Control the Reservoir Quality and Oiliness of Tight Sandstones? The Case of the Lower Cretaceous Quantou Formation in the Southern Songliao Basin, China. *Mar. Pet. Geol.* **2016**, *76*, 1–15. [CrossRef]

56. Zhang, L.; Lu, S.; Xiao, D.; Li, B. Pore Structure Characteristics of Tight Sandstones in the Northern Songliao Basin, China. *Mar. Pet. Geol.* **2017**, *88*, 170–180. [CrossRef]

57. Lundegard, P.D. Sandstone Porosity Loss: A "Big Picture" View of the Importance of Compaction. *J. Sediment. Res.* **1992**, *62*, 250–260. [CrossRef]

58. Ramm, M. Porosity-Depth Trends in Reservoir Sandstones: Theoretical Models Related to Jurassic Sandstones Offshore Norway. *Mar. Pet. Geol.* **1992**, *9*, 553–567. [CrossRef]

59. Holland, H.D.; Mackenzie, F.T.; Turekian, K.K. *Sediments, Diagenesis, and Sedimentary Rocks: Treatise on Geochemistry*, 2nd ed.; Elsevier Science: Amsterdam, The Netherlands, 2005; Volume 7, ISBN 978-0-08-044849-7.

60. Xi, K.; Cao, Y.; Jahren, J.; Zhu, R.; Bjørlykke, K.; Zhang, X.; Cai, L.; Hellevang, H. Quartz Cement and Its Origin in Tight Sandstone Reservoirs of the Cretaceous Quantou Formation in the Southern Songliao Basin, China. *Mar. Pet. Geol.* **2015**, *66*, 748–763. [CrossRef]

61. Gier, S.; Worden, R.H.; Krois, P. Comparing Clay Mineral Diagenesis in Interbedded Sandstones and Mudstones, Vienna Basin, Austria. *Geol. Soc. Lond. Spec. Publ.* **2018**, *435*, 389–403. [CrossRef]

62. Mozley, P.S.; Heath, J.E.; Dewers, T.A.; Bauer, S.J. Origin and Heterogeneity of Pore Sizes in the Mount Simon Sandstone and Eau Claire Formation: Implications for Multiphase Fluid Flow. *Geosphere* **2016**, *12*, 1341–1361. [CrossRef]

63. Salem, A.M.; Ketzer, J.M.; Morad, S.; Rizk, R.R.; Al-Aasm, I.S. Diagenesis and Reservoir-Quality Evolution of Incised-Valley Sandstones: Evidence from the Abu Madi Gas Reservoirs (Upper Miocene), the Nile Delta Basin, Egypt. *J. Sediment. Res.* **2005**, *75*, 572–584. [CrossRef]

64. Xi, K.; Cao, Y.; Liu, K.; Wu, S.; Yuan, G.; Zhu, R.; Kashif, M.; Zhao, Y. Diagenesis of Tight Sandstone Reservoirs in the Upper Triassic Yanchang Formation, Southwestern Ordos Basin, China. *Mar. Pet. Geol.* **2019**, *99*, 548–562. [CrossRef]

65. Jiu, B.; Huang, W.; Li, Y.; He, M. Influence of Clay Minerals and Cementation on Pore Throat of Tight Sandstone Gas Reservoir in the Eastern Ordos Basin, China. *J. Nat. Gas. Sci. Eng.* **2021**, *87*, 103762. [CrossRef]

66. Yuan, G.; Gluyas, J.; Cao, Y.; Oxtoby, N.H.; Jia, Z.; Wang, Y.; Xi, K.; Li, X. Diagenesis and Reservoir Quality Evolution of the Eocene Sandstones in the Northern Dongying Sag, Bohai Bay Basin, East China. *Mar. Pet. Geol.* **2015**, *62*, 77–89. [CrossRef]

67. Vincent, B.; van Buchem, F.S.P.; Bulot, L.G.; Jalali, M.; Swennen, R.; Hosseini, A.S.; Baghbani, D. Depositional Sequences, Diagenesis and Structural Control of the Albian to Turonian Carbonate Platform Systems in Coastal Fars (SW Iran). *Mar. Pet. Geol.* **2015**, *63*, 46–67. [CrossRef]

68. Wang, Q.; Xu, H.; Tang, D.; Yang, S.; Wang, G.; Ren, P.; Dong, W.; Guo, J. Indication of Origin and Distribution of Coalbed Gas from Stable Isotopes of Gas and Coproduced Water in Fukang Area of Junggar Basin, China. *AAPG Bull.* **2022**, *106*, 387–407. [CrossRef]

69. Lai, J.; Wang, G.; Ran, Y.; Zhou, Z.; Cui, Y. Impact of Diagenesis on the Reservoir Quality of Tight Oil Sandstones: The Case of Upper Triassic Yanchang Formation Chang 7 Oil Layers in Ordos Basin, China. *J. Pet. Sci. Eng.* **2016**, *145*, 54–65. [CrossRef]

70. Bjørlykke, K. *Petroleum Geoscience: From Sedimentary Environments to Rock Physics*; Springer Science & Business Media: Berlin/Heidelberg, Germany, 2010; ISBN 978-3-642-02332-3.

71. Dias, R.F.; Freeman, K.H.; Lewan, M.D.; Franks, S.G. $\Delta 13C$ of Low-Molecular-Weight Organic Acids Generated by the Hydrous Pyrolysis of Oil-Prone Source Rocks. *Geochim. Cosmochim. Acta* **2002**, *66*, 2755–2769. [CrossRef]

72. Schmid, S.; Worden, R.H.; Fisher, Q.J. Diagenesis and Reservoir Quality of the Sherwood Sandstone (Triassic), Corrib Field, Slyne Basin, West of Ireland. *Mar. Pet. Geol.* **2004**, *21*, 299–315. [CrossRef]

73. Freiburg, J.T.; Ritzi, R.W.; Kehoe, K.S. Depositional and Diagenetic Controls on Anomalously High Porosity within a Deeply Buried CO_2 Storage Reservoir—The Cambrian Mt. Simon Sandstone, Illinois Basin, USA. *Int. J. Greenh. Gas. Control.* **2016**, *55*, 42–54. [CrossRef]

74. Ehrenberg, S.N. Relationship Between Diagenesis and Reservoir Quality in Sandstones of the Garn Formation, Haltenbanken, Mid-Norwegian Continental Shelf1. *AAPG Bull.* **1990**, *74*, 1538–1558. [CrossRef]

75. Sborne, M.J.; Swarbrick, R.E. Diagenesis in North Sea HPHT Clastic Reservoirs—Consequences for Porosity and Overpressure Prediction. *Mar. Pet. Geol.* **1999**, *16*, 337–353. [CrossRef]

76. Blatt, H. Diagenetic Processes in Sandstones. In *Aspects of Diagenesis*; Scholle, P.A., Schluger, P.R., Eds.; SEPM Society for Sedimentary Geology, Special Publication: Tulsa, OK, USA, 1979; pp. 141–157.

77. McBride, E.F. Quartz Cement in Sandstones: A Review. *Earth Sci. Rev.* **1989**, *26*, 69–112. [CrossRef]

78. Yan, J.; Liu, C.; Zhang, W.; Li, B. Diagenetic Characteristics of the Lower Porosity and Permeability Sandstones of the Upper Paleozoic in the South of Ordos Basin. *Acta Geol. Sin.* **2010**, *84*, 272–279.

79. Kantorowicz, J.D. The Influence of Variations in Illite Morphology on the Permeability of Middle Jurassic Brent Group Sandstones, Cormorant Field, UK North Sea. *Mar. Pet. Geol.* **1990**, *7*, 66–74. [CrossRef]

80. Gaupp, R.; Matter, A.; Platt, J.; Ramseyer, K.; Walzebuck, J. Diagenesis and Fluid Evolution of Deeply Buried Permian (Rotliegende) Gas Reservoirs, Northwest Germany1. *AAPG Bull.* **1993**, *77*, 1111–1128. [CrossRef]

81. Towe, K.M. Clay Mineral Diagenesis as a Possible Source of Silica Cement in Sedimentary Rocks. *J. Sediment. Res.* **1962**, *32*, 26–28. [CrossRef]

82. Boles, J.R.; Franks, S.G. Clay Diagenesis in Wilcox Sandstones of Southwest Texas; Implications of Smectite Diagenesis on Sandstone Cementation. *J. Sediment. Res.* **1979**, *49*, 55–70. [CrossRef]

83. Ronald, C.S.; Crossey, L.J.; Hagen, E.S.; Heasler, H.P. Organic-Inorganic Interactions and Sandstone Diagenesis1. *AAPG Bull.* **1989**, *73*, 1–23. [CrossRef]

84. Tao, S.; Gao, X.; Li, C.; Zeng, J.; Zhang, X.; Yang, C.; Zhang, J.; Gong, Y. The Experimental Modeling of Gas Percolation Mechanisms in a Coal-Measure Tight Sandstone Reservoir: A Case Study on the Coal-Measure Tight Sandstone Gas in the Upper Triassic Xujiahe Formation, Sichuan Basin, China. *J. Nat. Gas. Geosci.* **2016**, *1*, 445–455. [CrossRef]

85. Rahman, M.J.J.; Worden, R.H. Diagenesis and Its Impact on the Reservoir Quality of Miocene Sandstones (Surma Group) from the Bengal Basin, Bangladesh. *Mar. Pet. Geol.* **2016**, *77*, 898–915. [CrossRef]

86. Grigsby, J.D. Origin and Growth Mechanism of Authigenic Chlorite in Sandstones of the Lower Vicksburg Formation, South Texas. *J. Sediment. Res.* **2001**, *71*, 27–36. [CrossRef]

87. Virolle, M.; Brigaud, B.; Luby, S.; Portier, E.; Féniès, H.; Bourillot, R.; Patrier, P.; Beaufort, D. Influence of Sedimentation and Detrital Clay Grain Coats on Chloritized Sandstone Reservoir Qualities: Insights from Comparisons between Ancient Tidal Heterolithic Sandstones and a Modern Estuarine System. *Mar. Pet. Geol.* **2019**, *107*, 163–184. [CrossRef]

88. Yao, J.L.; Wang, Q.; Zhang, R.; Li, S.T. Forming Mechanism and Their Environmental Implications of Chlorite-Coatings in Chang 6 Sandstone (Upper Triassic) of Hua-Qing Area, Ordos Basin. *Acta Sedimentol. Sin.* **2011**, *29*, 72–79.

89. Midtbø, R.E.A.; Rykkje, J.M.; Ramm, M. Deep Burial Diagenesis and Reservoir Quality along the Eastern Flank of the Viking Graben. Evidence for Illitization and Quartz Cementation after Hydrocarbon Emplacement. *Clay Miner.* **2000**, *35*, 227–237. [CrossRef]

90. Baker, J.C. Diagenesis and Reservoir Quality of the Aldebaran Sandstone, Denison Trough, East-Central Queensland, Australia. *Sedimentology* **1991**, *38*, 819–838. [CrossRef]

91. Littke, R.; Zieger, L. Deposition, Diagenesis and Petroleum Generation Potential of Pennsylvanian Coals and Coal-Bearing Strata in Western Germany: A Review. *Z. Der Dtsch. Ges. Für Geowiss.* **2019**, *170*, 289–309. [CrossRef]

92. Zhou, Y.; Ji, Y.; Xu, L.; Che, S.; Niu, X.; Wan, L.; Zhou, Y.; Li, Z.; You, Y. Controls on Reservoir Heterogeneity of Tight Sand Oil Reservoirs in Upper Triassic Yanchang Formation in Longdong Area, Southwest Ordos Basin, China: Implications for Reservoir Quality Prediction and Oil Accumulation. *Mar. Pet. Geol.* **2016**, *78*, 110–135. [CrossRef]

93. Nguyen, D.; Horton, R.A.; Kaess, A.B. Diagenesis of the Oligocene Vedder Formation, Greeley Oil Field, Southern San Joaquin Basin, California. In Proceedings of the Pacific Section AAPG, SPE and SEPM Joint Technical Conference, Bakersfield, CA, USA, 29 April 2014.

94. Beard, D.C.; Weyl, P.K. Influence of Texture on Porosity and Permeability of Unconsolidated Sand1. *AAPG Bull.* **1973**, *57*, 349–369. [CrossRef]

95. Ramm, M. Reservoir Quality and Its Relationship to Facies and Provenance in Middle to Upper Jurassic Sequences, Northeastern North Sea. *Clay Miner.* **2000**, *35*, 77–94. [CrossRef]

96. Rezaee, R. *Fundamentals of Gas Shale Reservoirs*; John Wiley & Sons: Hoboken, NJ, USA, 2015; ISBN 978-1-118-64579-6.

97. Ehrenberg, S.N. Influence of Depositional Sand Quality and Diagenesis on Porosity and Permeability; Examples from Brent Group Reservoirs, Northern North Sea. *J. Sediment. Res.* **1997**, *67*, 197–211. [CrossRef]

98. Lai, J.; Wang, G.; Ran, Y.; Zhou, Z. Predictive Distribution of High-Quality Reservoirs of Tight Gas Sandstones by Linking Diagenesis to Depositional Facies: Evidence from Xu-2 Sandstones in the Penglai Area of the Central Sichuan Basin, China. *J. Nat. Gas. Sci. Eng.* **2015**, *23*, 97–111. [CrossRef]

99. Fisher, Q.J.; Casey, M.; Clennell, M.B.; Knipe, R.J. Mechanical Compaction of Deeply Buried Sandstones of the North Sea. *Mar. Pet. Geol.* **1999**, *16*, 605–618. [CrossRef]

100. Rossi, C.; Marfil, R.; Ramseyer, K.; Permanyer, A. Facies-Related Diagenesis and Multiphase Siderite Cementation and Dissolution in the Reservoir Sandstones of the Khatatba Formation, Egypt's Western Desert. *J. Sediment. Res.* **2001**, *71*, 459–472. [CrossRef]

101. Zhu, R.; Zou, C.; Zhang, N.; Wang, X.; Cheng, R.; Liu, L.; Zhou, C.; Song, L. Diagenetic Fluids Evolution and Genetic Mechanism of Tight Sandstone Gas Reservoirs in Upper Triassic Xujiahe Formation in Sichuan Basin, China. *Sci. China Ser. D-Earth Sci.* **2008**, *51*, 1340–1353. [CrossRef]

102. Zou, C.; Zhu, R.; Liu, K.; Su, L.; Bai, B.; Zhang, X.; Yuan, X.; Wang, J. Tight Gas Sandstone Reservoirs in China: Characteristics and Recognition Criteria. *J. Pet. Sci. Eng.* **2012**, *88–89*, 82–91. [CrossRef]

103. Ji, Y.; Zhou, Y.; Liu, Y.; Lu, H.; Liu, Q.; Wang, Y. The Impact of Sedimentary Characteristics on the Diagenesis and Reservoir Quality of the 1st Member of Paleogene Funing Formation in the Gaoyou Subbasin. *Acta Geol. Sin.* **2014**, *88*, 1299–1310.

104. Wu, H.; Ji, Y.; Liu, R.; Zhang, C.; Chen, S. Insight into the Pore Structure of Tight Gas Sandstones: A Case Study in the Ordos Basin, NW China. *Energy Fuels* **2017**, *31*, 13159–13178. [CrossRef]

Article

Source-to-Sink Comparative Study between Gas Reservoirs of the Ledong Submarine Channel and the Dongfang Submarine Fan in the Yinggehai Basin, South China Sea

Yue Yao [1], Qiulei Guo [2],* and Hua Wang [3]

[1] College of Marine Science and Technology, China University of Geosciences, Wuhan 430074, China; missstaru@foxmail.com
[2] School of Earth and Space Sciences, Peking University, Beijing 100871, China
[3] School of Earth Resources, China University of Geosciences, Wuhan 430074, China; wanghua@cug.edu.cn
* Correspondence: qiulei.guo@pku.edu.cn

Abstract: The Ledong submarine channel and the Dongfang submarine fan, two remarkable sedimentary systems developed during the late Miocene, are considered promising hydrocarbon reservoirs in the Yinggehai Basin of the South China Sea. A comparative study was conducted to reveal the differences between the source-to-sink characteristics of the two gas-bearing and gravity-driven depositional systems to determine their provenances, formation mechanisms and migration paths as well as their key controlling factors. The heavy mineral assemblages and detrital zircon U-Pb dating results suggest that the Ledong channel was fed by the Hainan provenance from the eastern margin, whereas the Dongfang fan was supplied by northwestern terrigenous sources. The relative sea level transgression and sufficient sediment supply triggered the delivery of deltaic loads toward the continental shelves. Seismic data show that fracture activity had a great impact on the tectono-morphologic features of the margins. During downward flow, the gravity flow along the Yingdong Slope encountered steeply falling faulted slope break belts and formed the Ledong incised channel, and the gravity flow of the Yingxi Slope moved through the gently dipping flexural break slope zone and formed the Dongfang dispersed lobe deposits. Since ca. 30 Ma, the sedimentary center has been migrating from the north to the southeast, which produced a clear control of the southeastward distribution pattern of these two sedimentary systems. Observations of cores and thin sections indicate that the rock structures and their compositions are more mature in the Dongfang channel than in the Ledong fan. This study documents significant differences and similarities by comparing the source-to-sink processes of the two gravity-driven systems that developed in the Yinggehai Basin and provides analogies for understanding similar submarine sedimentary systems that developed under similar geological contexts worldwide.

Keywords: submarine channel; submarine fan; source-to-sink; gas reservoirs; South China Sea; Yinggehai Basin

Citation: Yao, Y.; Guo, Q.; Wang, H. Source-to-Sink Comparative Study between Gas Reservoirs of the Ledong Submarine Channel and the Dongfang Submarine Fan in the Yinggehai Basin, South China Sea. *Energies* **2022**, *15*, 4298. https://doi.org/10.3390/en15124298

Academic Editor: Reza Rezaee

Received: 7 May 2022
Accepted: 7 June 2022
Published: 11 June 2022

Publisher's Note: MDPI stays neutral with regard to jurisdictional claims in published maps and institutional affiliations.

1. Introduction

Submarine fans and channels are prominent marine gravity flow sedimentary features; they compose the largest sediment accumulations on the continental margin and seafloor, which may be prolific hosts for oil and gas resources [1–5]. Worldwide, gravity flow systems were extensively developed during the Miocene icehouse periods, particularly in the basins of the South China Sea [6,7]. Some examples of Miocene submarine channels and fans are from the Pearl River mouth slope, the eastern Qiongdongnan margins, the Santos Basin and the south Greenland margin [1,2,8,9]. Both the Dongfang submarine fan and the newly discovered Ledong submarine channel, which developed in the late Miocene, are two important sweet spot zones in the Yinggehai Basin of the South China Sea, where numerous large gas fields have successively been discovered [10,11]. After years of natural gas exploration, more than 260 billion cubic meters of proven natural gas reserves have been confirmed in the Dongfang and Ledong districts [12,13]. Similar to many terrestrial, lacustrine and marine basins around the world, the Yinggehai Basin is a multisource supply basin with diverse surrounding river systems and complex lithologic distributions [14–16]. The two sedimentary systems present very different characteristics with respect to morphology, provenance, transport pathways and sedimentary characteristics. Therefore, it is of great significance and a great necessity to conduct a comparative study of the "source-route-sink" processes of these two typical gravity flow sedimentary systems.

The concept of source-to-sink links together all of the steps from sediment formation to final deposition, with regard to the denudation of sources, the transfer of erosion products and the ultimate sedimentation process [17–20]. Previous researchers have discussed the source-to-sink system of the Yinggehai Basin and have confirmed two major sources of the basin, one being the Yangtze and Indosinian plates in the northwestern part of the basin and the other being the Hainan Island uplift in the eastern part of the basin [21–24]. However, little attention has been given to the similarities, differences and correlations of the source-to-sink processes among different submarine sedimentary systems. Specifically, no systematic source-to-sink comparative study on the Ledong gravity flow channel system and Dongfang submarine fan system has been conducted. Hence, it is necessary to strengthen the comparison and summarize the regularity of sediments under different source and sink modes. Scholars have conducted detailed studies of the sedimentary system developed in the Ledong and Dongfang districts [25–27]. For example, Huang et al. (2021) [25] studied the depositional facies, logging facies and petrographic characteristics of the submarine fan developed in the Dongfang district. Li et al. (2021) [12] studied the filling evolution characteristics of axial gravity flow channels in the Ledong district. Nonetheless, the controlling agents for the differences in the sedimentary systems in the submarine fans, which dominate the Dongfang area, and the gravity flow channels, which dominate the Ledong area, are still under a dense fog. More work is needed to explain the differences in the material sources and migration processes of the sediments between the two and to reveal the dissimilarities in their formation mechanisms.

This study aims to systematically compare and clarify the differences and develop correlations between the "source–sink" processes of these two typical gas-bearing and gravity-driven sedimentary systems that developed in the Miocene Huangliu Formation in the Ledong and Dongfang districts of the Yinggehai Basin with macroscopic to microscopic petrological analyses, heavy mineral assemblages, zircon U-Pb dating, combined well logging and seismic analyses. Specifically, from the source–sink perspective, the potential origin, triggering mechanism, transport path and depositional characteristics have been investigated. The study discusses why two different types of submarine gravity flow systems developed in the same basin during the same geological time and whether there was a key controlling factor influencing the development and distribution patterns of the channels and lobes. This study aims to develop some valuable geological guidance for future explorations for gas reservoirs in the South China Sea and to be an important reference to allow investigators to predict the evolution of other comparable sedimentary systems in analogous depositional environments.

2. Geological Setting

2.1. The Yinggehai Basin

The Yinggehai Basin (Figure 1a) in the South China Sea is an important Cenozoic hydrocarbon-bearing basin located on the northwestern margin of the South China Sea between Hainan Island and the Indo-China Peninsula [28–30]. Unlike the Beibuwan, Qiongdongnan and Zhujiangkou basins in the western part of the South China Sea, the Yinggehai Basin has a unique NNW–SSE trend, with an area of 12.17×10^4 km^2 [31–33]. The Yinggehai Basin is a transformation extensional basin formed in the early Eocene by the superposition of various tectonic movements, such as the subduction of the Indian plate into the Eurasian plate, the uplift of the Qinghai-Tibet orogen and the expansion of the South China seafloor [24,30]. The basin features a high geothermal gradient (4 to 6 °C/100 m), a large-scale anomalous pressure coefficient (1.0 to 2.3), rapid subsidence (500 to 1400 m/Ma) and mud diapering [34,35].

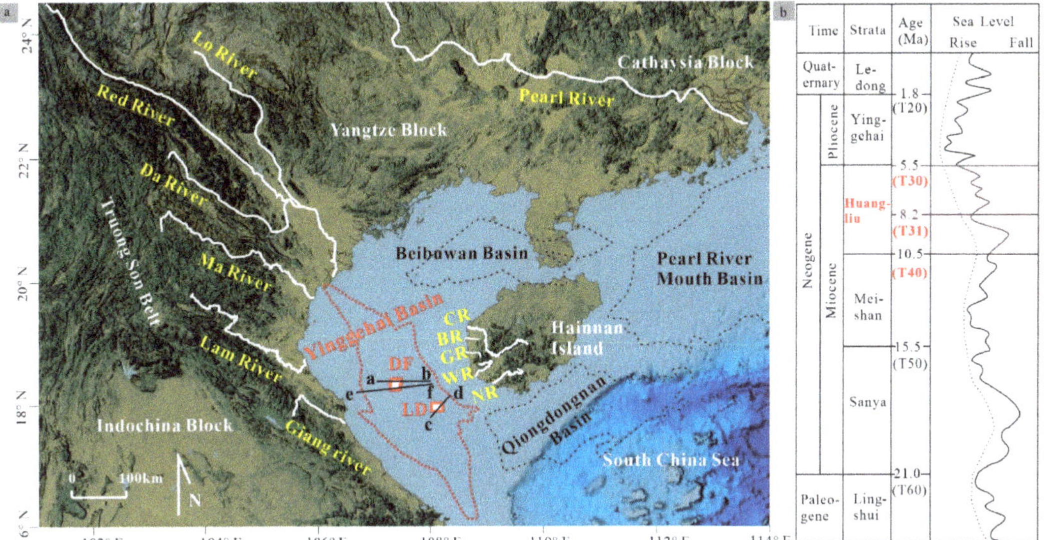

Figure 1. (**a**) Topographic map showing the location of the Yinggehai Basin, which is in the northwestern part of the South China Sea, and the adjacent continental blocks, sedimentary basins and major drainage systems. DF: Dongfang, LD: Ledong, NR: Ningyuan River, WR: Wanglou River, GR: Ganen River, BR: Beili River, CR: Changhua River. (a–b, c–d, e–f are seismic lines). (**b**) Stratigraphic chart with sea-level curve (modified from Huang et al. (2020) [11]). T20, T30, T40, T50 and T60 represent the seismic boundaries between the different formations, and the studied strata of the Huangliu Formation are highlighted in the chart. T30 and T40 are the top and bottom surface layers of the Huangliu Formation, respectively, whereas T31 divides the Huangliu Formation into the upper member and lower member.

At present, a relatively complete Cenozoic stratum has been established in the Yinggehai Basin. The stratigraphy from Paleogene to Quaternary are the Lingshui, Sanya, Meishan, Huangliu, Yinggehai and Ledong Formations (Figure 1b) [36,37]. Among these, the late Miocene Huangliu Formation (10.5–5.5 Ma) is located between the basin-wide unconformities, namely, the T40 and T30 seismic boundaries, which are further divided into upper and lower segments by the T31 surface, dated at 8.2 Ma (Figure 1b).

The Yinggehai Basin is divided into three tectonic units (Figure 2), which are the Yingxi Slope, the Central Depression and the Yingdong Slope, by the Yingxi Fault zone in the west and the No. 1 Fault zone in the east [28,30]. The Yingxi Fault zone is subdivided into the Da, Ma, Lam, Troung Son faults, etc. The No. 1 Fault zone is subdivided into the Yingdong Fault and the No. 1 Fault (Figure 2) [38,39]. The fault system in the basin is complex, and the trends of these fault zones are mainly SSE–SE, among which the No. 1 Fault zone on the east side of the basin is dominant in the basin and is a large basin-controlling fault that developed in a SE trend.

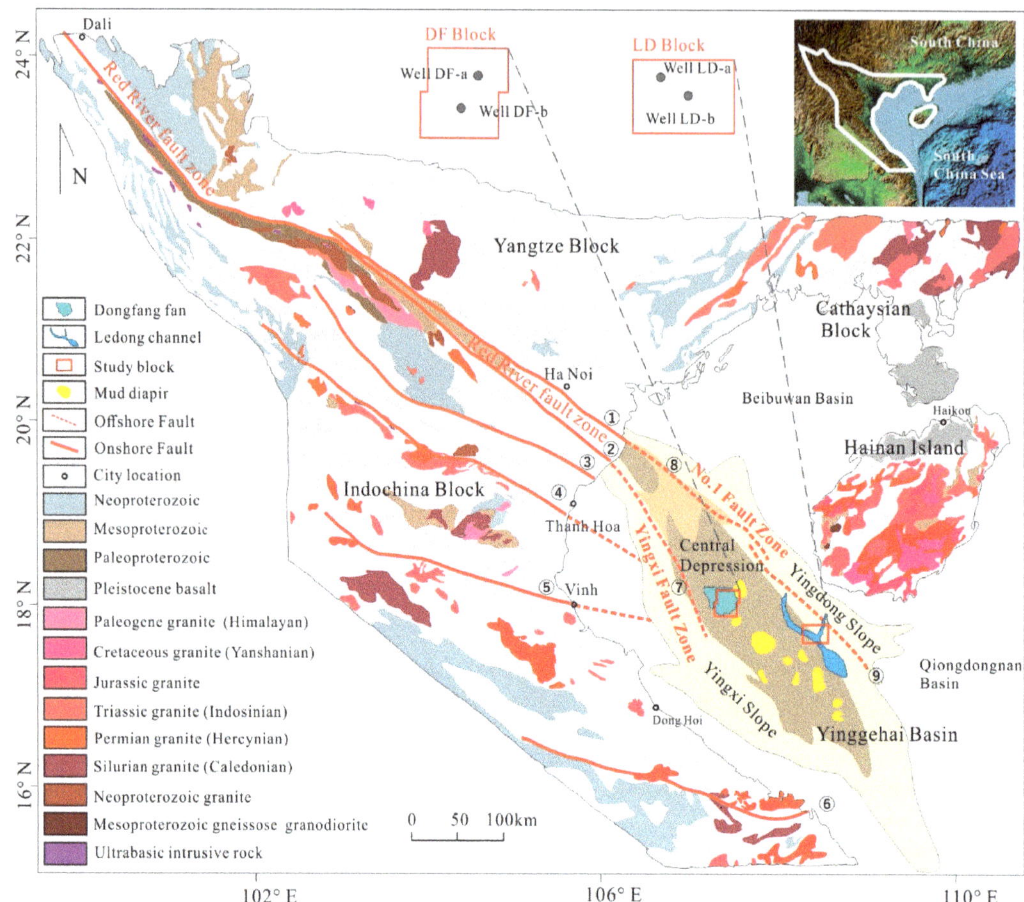

Figure 2. Igneous and Proterozoic rocks are distributed around the Yinggehai Basin. The Dongfang Block and submarine fan of this study are located in the Central Depression, and the Ledong Block and submarine channel are located beside the Yingdong Slope. Some major onshore faults are shown, which include the ① Lo River Fault, ② Red River Fault, ③ Da River Fault, ④ Ma River Fault, ⑤ Lam River Fault, ⑥ Troung Son Fault, ⑦ Yingxi Fault, ⑧ Yingdong Fault and ⑨ No. 1 Fault. The No. 1 Fault zone comprises the Yingfong Fault and the No. 1 Fault. The small map in the upper right corner shows the regional location of the geological map.

The Yinggehai Basin had sufficient material sources from the numerous nearby river systems (Figure 1a). The provenance mainly consists of the Red River source from the Yangtze Plate in the northern part of the Yinggehai Basin, the Truong Son Belt source along the coast of Vietnam in the west and the Hainan source from Hainan Island in the east (Figure 1a) [40–42]. The Truong Son Belt and the Red River source are collectively referred to as the northwestern provenance. The current northwestern provenance river system is relatively large, among which the Red River drainage includes the Red River, Da River and Lo River, and the Truong Son Belt drainage includes the Ma River, Lam River and Giang River (Figure 1a). There are many river systems along the west coast of Hainan Island, including the Changhua River, Beili River, Ganen River, Ningyuan River and Wanglou River.

The Truong Son Belt developed in the eastern margin of Vietnam and spreads in an "S" pattern from north to south. Due to high gravitational potential energy, the sand contents of Vietnam rivers are generally high [43,44]. The northern part of the basin is dominated by Red River source input. Moreover, after its upper reaches were captured by the Yangtze River (before the early Miocene), the Miocene source area shrunk somewhat from its previous extent, but its source supply capacity was still large [7]. The Hainan Island Uplift was formed in the early Cretaceous period and is mainly located in the central and southern parts of Hainan Island, with east–west fault zones on both sides, covering an area of 3.44×10^4 km^2, and the current height of Wuzhishan, its main peak, is approximately 1887 m [45,46].

2.2. The Dongfang Submarine Fan and Ledong Submarine Channel

Exploration practice has shown that the Huangliu Formation in the Yinggehai Basin is the most developed section of gravity flow [10,25]. The sea level dropped (Figure 1b) during this period, and the coastline migrated toward the basin, which was conducive to the long-distance transport of clastic material to be deposited in the depression.

The Dongfang Block (Figure 2) is located in the north-central part of the Central Depression of the Yinggehai Basin, near the central mud diapir anticline structure, with a total area of approximately 2400 km^2 [6,23]. Multiple submarine gravity flow fan sedimentary systems developed in the Huangliu Formation of the Dongfang district, and the submarine fan of this study is located in the Dongfang Block (Figure 2). The submarine fan covers an area of approximately 700 km^2 and is distributed in a NW–SE orientation. It developed in the first member of the upper Miocene Huangliu Formation between seismic interfaces T30 and T31 (Figure 1b).

The Ledong Block (Figure 2) is located between the Yingdong Slope belt and the Central Depression belt of the Yinggehai Basin in the western continental shelf of the northern part of the South China Sea [26,47]. This block is currently in the exploration evaluation stage, and the main gas-bearing zone is the Huangliu Formation. The submarine gravity flow channel is located in the second member of the Upper Miocene Huangliu Formation, which is located between the seismic interface T31-T40 (Figure 1b), with a depth of 3800~4500 m, which was deposited in a shallow sea sedimentary environment. The channel consists of two branches: the north branch and the east branch (Figure 2). The east branch developed parallel to the Yingdong slope, and the north branch is perpendicular to the slope. Overall, the channel is approximately 190 km in length, 2–7.2 km in width and 80–550 m in depth. The northern and southern branches intersect on the slope and continue to develop axially toward the southeast for nearly 20 km before they begin to disperse, with the tail disappearing in a fan-like pattern in the combination zone of the Yinggehai and Qiongdongnan basins.

3. Data and Method

There is currently a total of 23 wells drilled in the Ledong and Dongfang study areas. However, due to the difficulty and limitations of coring in offshore engineering projects, there are currently only four coring wells in the study area. Through this study, the well logging curves, cores, rock cuttings and rock slice samples from the four most typical wells in the Dongfang and Ledong districts were selected for comparative analysis, including Well DF-a and Well DF-b drilled in the Dongfang submarine fan and Well LD-a and Well LD-b drilled in the Ledong submarine channel (Figure 2).

During the field investigations, five river estuary sand samples were collected on the west side of Hainan Island from the Ningyuan River, Wanglou River, Beili River, Ganen River and Changhua River (Figure 1). Data from the Red River and eastern Vietnam River groups riverbank sand samples were gathered from Fyhn's published reports [44]. Eight core samples were selected for heavy mineral composition analysis, and four samples among them were selected for zircon U-Pb dating analysis. Details about the data for zircon dating and heavy mineral analyses are listed in Table 1. The seismic data for the Ledong and Dongfang districts were provided by the China National Offshore Oil Corporation (Hainan Branch). The poststack 3D seismic data, having a bin size of 12.5 × 12.5 m^2 and a full fold value of 48, used in this study were acquired between 2001 and 2015. Datasets were characterized by an effective frequency range of 15–55 Hz, with a predominant frequency of approximately 40 Hz.

Table 1. Zircon and heavy mineral sample details.

Sample	Name	Origination	Depth	Location	Formation
Dongfang Sample	DF-a	Borehole rock cuttings	2876 m	Dongfang fan	Huangliu T31-T30
	DF-a-2	Borehole rock cuttings	2832 m	Dongfang fan	Huangliu T31-T30
	DF-b	Borehole rock cuttings	3120 m	Dongfang fan	Huangliu T31-T30
	DF-b-2	Borehole rock cuttings	3156 m	Dongfang fan	Huangliu T31-T30
Ledong Sample	LD-a	Borehole rock cuttings	4104 m	Ledong channel	Huangliu T40-T31
	LD-a-2	Borehole rock cuttings	4180 m	Ledong channel	Huangliu T40-T31
	LD-b	Borehole rock cuttings	4072 m	Ledong channel	Huangliu T40-T31
	LD-b-2	Borehole rock cuttings	4158 m	Ledong channel	Huangliu T40-T31
Hainan Island Sample	NR	Nngyuan River sand	Surface	Hainan estuary	Quaternary
	WR	Wanglou River sand	Surface	Hainan estuary	Quaternary
	CR	Changhua River sand	Surface	Hainan estuary	Quaternary
	BR	Beili River sand	Surface	Hainan estuary	Quaternary
	GR	Ganen River sand	Surface	Hainan estuary	Quaternary
Northwest Provenance Sample	Red	Red River sand	Surface	Riverbank	Quaternary
	Ma	Ma River sand	Surface	Riverbank	Quaternary
	Lam	Lam River sand	Surface	Riverbank	Quaternary
	Giang	Giang River sand	Surface	Riverbank	Quaternary

The detrital zircon U/Pb dating method, heavy mineral assemblage method, core and sandstone thin section observations and other regional geological data were combined to determine the rock compositions and potential detrital material source areas. Samples were sent to the Langfang Geological Service Co., Ltd., Langfang, China, for zircon and heavy mineral identification, separation and counting. In this study, ten types of heavy minerals with relatively high and comparable contents were screened from the samples, including extremely stable heavy minerals such as zircon, tourmaline and rutile; stable heavy minerals such as garnet, sphene, magnetite, leucoxene, anatase, monazite and staurolite; and unstable minerals such as chlorite and pyroxene [41,48]. The study calculated the relative weight percentages of the heavy minerals as well as the zircon–tourmaline–rutile (ZTR) index, which measures the maturity of heavy minerals [49]. Heavy minerals with great maturity (such as zircon, rutile or tourmaline) have strong resistance to weathering and can, essentially, maintain their original properties after long-distance transportation and screening [21]. In situ zircon U-Pb isotope dating was carried out at Wuhan Shangpu Analytical Technology Co., Ltd., Wuhan, China, by using an Agilent 700 Inductively Coupled Plasma Mass Spectrometer and a GeoLas HD coherent 193 nm excimer laser ablation system. The test used international standard material samples 91,500, GJ-1 and NIST610 as the isotope ratio calibration standard sample, the isotope ratio monitoring standard sample and the trace element calibration standard sample, respectively. The processing software used was ICPMSDATACAL10.8. U-Pb ages with concordance values greater than or equal to 90% were retained for zircon dating in the study.

Through the seismic preprocessing and interpretation method, the internal structure of the sedimentary system and the basin's tectono-morphologic characteristics were analyzed. Schlumberger's Geoframe software was applied to visualize the subsurface structure through a three-dimensional seismic reflection structure, amplitude strength and stratigraphic continuity profile. The faulting system, morphology of the continental shelf and gravity flow in the Ledong and Dongfang districts were interpreted and identified. Through the seismic layer flattening method, the seismic data volume was flattened along the seismic reflection layer corresponding to the specific geological interface, thereby reconstructing the ancient landform and restoring the paleogeomorphology of the Huangliu Formation. The thickness map and instantaneous amplitude map were based on seismic time-to-depth conversion and attribute extraction technology to analyze and display the distribution patterns of the channels and the lobes as well as lithology variation.

4. Results

4.1. Well Logging-Core-Thin Section Observation

In the Dongfang area, logging is mainly characterized by medium- to high-amplitude bell- or box-shaped curves that have positive grain sequence features (Figure 3a). The sandstone of the Huangliu Formation in Well DF-a is mainly fine to very fine sandstone with particle diameters distributed in the range of 0.120–0.040 mm, and the maximum thickness of a single sand body is 37 m. The core image (Figure 3a) shows light gray massive bedding and parallel bedding features. Based on the thin section analysis of Well DF-a sediment and other Dongfang sediment slices (Figure 3c), the detrital compositions consist of quartz (average 61%), potassium feldspar (average 3%) and other rock debris (average 9.5%), with medium sorting and subrounded roundness. The quartz comprises a large amount of monocrystalline quartz (55%) and a small amount of polycrystalline quartz (6%). The rock debris contains mainly metamorphic rocks (average 9%), and the content of the argillaceous matrix is relatively low (1%).

The well logging curves in the Ledong district are predominantly box-shaped, with an abrupt interface between the sandstone and mudstone at ca. 4045 m. (Figure 3b). The lithology of the Huangliu Formation in Well LD-b is mainly medium sandstone, with locally developed fine sandstone, and the maximum thickness of a single sand body can reach 36 m. Through core observation (Figure 3b), typical gravity flow sedimentary structures, such as massive bedding, Bauma sequences and mud gravel, can be identified. The massive sandstone contains apparent gravel and is poorly sorted. The results of the rock thin section analysis (Figure 3d) in Well LD-b show that the grain diameters are mainly distributed in the range of 0.500–0.050 mm, with poor sorting and subangular roundness. The average contents of monocrystalline quartz, polycrystalline quartz, potassium feldspar, plagioclase and rock debris are 58%, 20%, 7%, 3% and 1%, respectively. The rock debris is mainly composed of intrusive rocks.

The above logging-core-thin section observation (Figure 3) results present the most representative results among 23 logging curves, 160 m core and 129 thin sections. The results of the drilling and core data not shown in the paper are similar to the above given data.

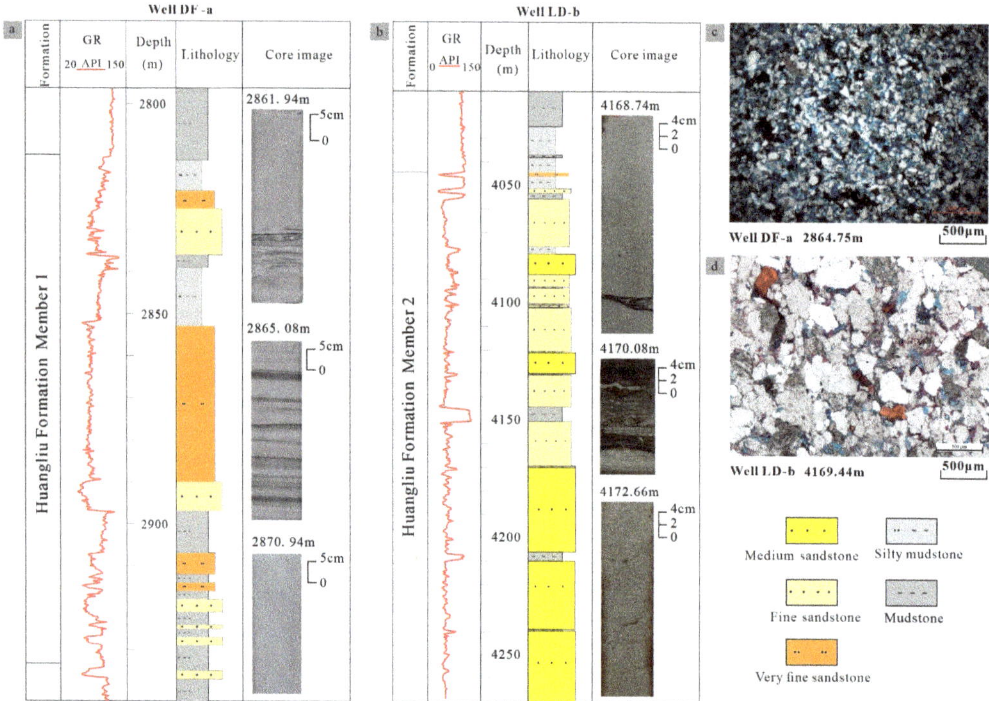

Figure 3. Well logging curves, lithology interpretation, core images and thin section observations: (**a**) Well DF-a in the Huangliu Formation upper member. (**b**) Well LD-b in the Huangliu Formation lower member. (**c**) Well DF-a sample thin section observation at 2864.75 m. (**d**) Well LD-b sample thin section observation at 4169.44 m.

4.2. Heavy Mineral Assemblage and Zircon U-Pb Dating

The heavy mineral assemblage results (Figure 4) of drilling samples from the Dongfang block demonstrate that leucoxene (average 65.85%), zircon (average 7.24%) and tourmaline (average 10.47%) dominate in the Dongfang wells, with an average ZTR index of 19.68. Drilling samples from the Ledong block wells display high contents of magnetite (30.54%), zircon (average 28.20%), tourmaline (average 14.65%) and garnet (average 11.73%), with an average ZTR index of 40.86. The fluvial sand samples from Hainan Island are dominated by zircon (average 27.73%), garnet (average 16.47%) and magnetite (average 20.44%), with an average ZTR index of 41.46. The river sands of eastern Vietnam and the Red River are dominated by leucoxene (average 42.33%), zircon (average 12.83%), rutile (average 15.56%), garnet (average 13.86%) and titanite (average 12.36%), with an average ZTR index of 28.39.

The zircon U-Pb dating spectra of both the Hainan river sand samples and the Ledong channel sediment samples (Figure 5) show bimodal characteristics, containing the main peaks of two geological ages: Cretaceous (97–99 Ma) and Permian–Triassic (232–249 Ma), respectively. The Cretaceous peaks are higher in the Ningyuan and Wanglou Rivers of Hainan Island, and the Permian–Triassic peaks are higher in the Changhua River of Hainan Island. The two peaks of the Ledong samples are similar in height, and some small Jurassic (151–153 Ma) peaks occur. The zircon dating patterns of the river sands, having northwestern provenance and the rock cuttings of the Huangliu Formation from the Dongfang submarine fan, display multipeak patterns. Among these, the samples from the Red River, Lam River, Ma River and the Dongfang block have distinct Paleogene (30–37 Ma), Permian–Triassic (243–262 Ma) and Silurian–Devonian (420–456 Ma) peaks, whereas the Giang River has significant Triassic peaks (248 Ma).

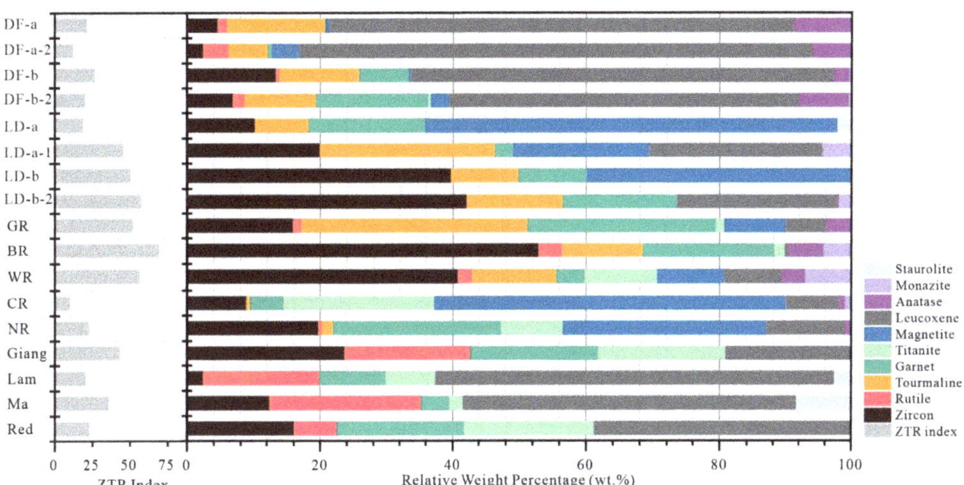

Figure 4. Heavy mineral assemblage and zircon–tourmaline–rutile (ZTR) index ratios of present-day fluvial sand samples, Ledong borehole samples and Dongfang borehole samples. The northwestern provenance data, such as that from the Giang River, Lam River, Ma River and Red River sands, were abstracted from Fyhn et al. (2019) [44].

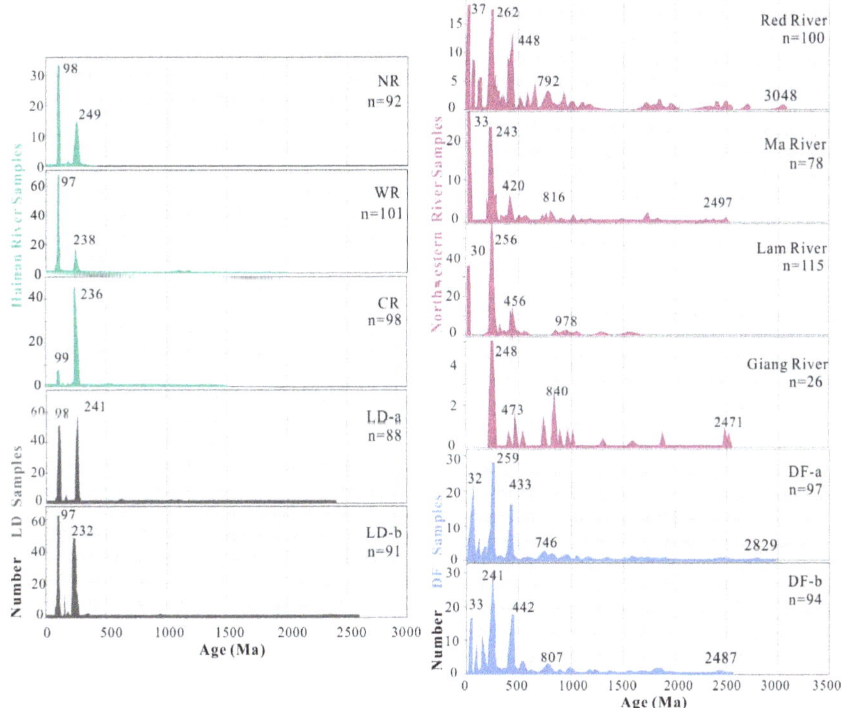

Figure 5. Detrital zircon Pb206/U238 age spectrum of Hainan River samples (green), Ledong samples (gray), northwestern river samples (purple) and Dongfang samples (blue). The "n" denotes the total number of zircons used for dating.

The result of the heavy mineral composition and zircon U-Pb geochronology of the selected four samples collected in this paper are typical and representative, which can provide a reliable basis for the study of sources in the two regions. Some scholars have previously made geochemical analyses on the sediments of the Huangliu Formation in the Dongfang district and Ledong district, and the findings reflected by the previous results are consistent with the results of the sampling analysis in this paper [21,22,36,37].

4.3. Seismic Processing

The seismic results show that the Dongfang fan developed in the northwest Central Depression in a flexure zone near the mud diapir, and it was distributed in a lenticular form between seismic interfaces T31 and T30 (Figure 6a). It is a composite submarine fan formed by the superposition of multistage lobes and channels. The topography of the Dongfang region in the late Miocene was relatively gentle and flat (Figure 6b), with slopes ranging between 2 to 3.2 degrees near well DF-b and 5.6 to 8 degrees near well DF-a.

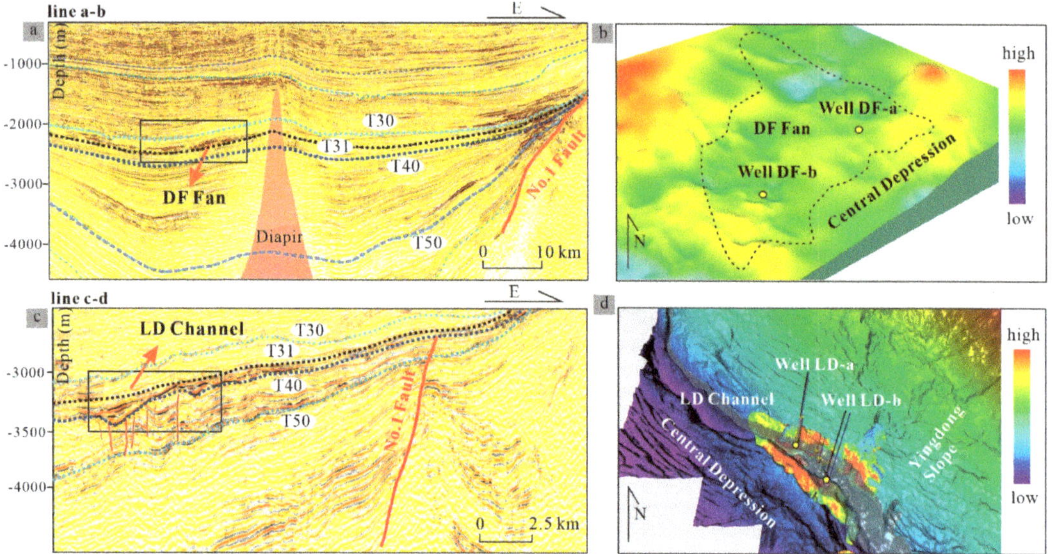

Figure 6. Seismic profiles and regional palaeotopography restorations (see the locations of lines a-b and c-d in Figure 1): (**a**) Seismic profile of the Dongfang fan developed in the second member of the Huangliu Formation between T30 and T31. (**b**) Dongfang Block paleotopography restoration. (**c**) Seismic profile of the Ledong channel developed in the first member of the Huangliu Formation between T31 and T40. (**d**) Ledong Block paleotopography restoration.

The Ledong channel, which developed under the No. 1 Fault slope break zone of the Yingdong Slope, is located between seismic interfaces T40 and T31 (Figure 6c). The bottom interface of the channel is U-shaped or V-shaped, with strong amplitude and high continuity. Several small hidden faults developed at the channel bottom. The paleogeomorphic restoration shows that, in the late Miocene Huangliu Formation, the terrain from Yingdong Slope to the Central Depression was steep (Figure 6d), with slopes near Well LD-a and Well LD-b that were both greater than 15 degrees.

According to the seismic reflection interfaces, the contact relationship between the sedimentary strata and the distribution of the sand layer, the submarine channel and the submarine fan sand body are divided into two stages. The first-stage sand body is located at the bottom of the gravity flow and was formed earlier, whereas the second-stage sand body is located above the first-stage sand body and was formed later. The sand thickness map shows that, in the early stage of submarine fan development (Figure 7a), the reservoir mainly developed on the southwestern side, and the average thickness of the lobe is 38 m. In the later stage of the development of the submarine fan (Figure 7b), the fan area retreated to the northwest with decreased average thickness (30 m) and more evident channelization. The instantaneous amplitude map (Figure 7c,d) shows that the Ledong channel consists of two upstream tributaries in the northwest and northeast and one downstream tributary in the southeast, distributed in a "Y" shape. The sand reservoir in the first stage of channel development shows high continuity and connectivity along the channel. In the second stage of channel development, the sediment mud content increased, showing a strip with a small width on the plane.

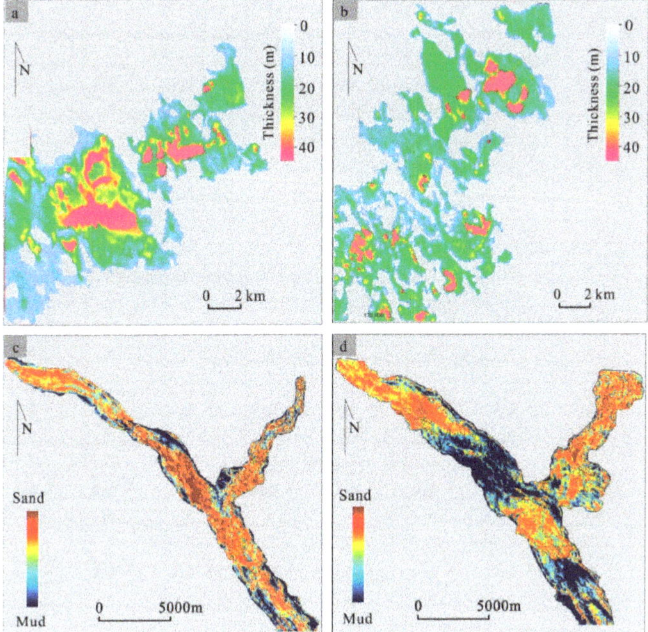

Figure 7. Thickness maps and Instantaneous amplitude maps: (**a**) Thickness map of the Dongfang submarine fan developed in the first stage. (**b**) Thickness map of the Dongfang submarine fan developed in the second stage. (**c**) Instantaneous amplitude map of the Ledong submarine channel developed in the first stage. (**d**) Instantaneous amplitude map of the Ledong submarine channel developed in the second stage.

5. Discussions

5.1. Provenance Comparison

The combination of data from logging, cores, thin-section microscopic observations, heavy mineral analyses and zircon dating indicates that the provenances of the studied areas have different sources. Specifically, the Dongfang submarine fan sedimentary system developed in the Miocene Huangliu Formation is from the northwestern margin of the Yinggehai Basin, and the source of the Ledong channel sedimentary system, which developed in the same geological period, is from the eastern margin of the basin.

The maturity of the rock structure in the Ledong area is higher than that in the Dongfang area, suggesting that the transportation distance of provenance in the Dongfang area is longer than that in the Ledong area. The clastic particles from the Dongfang Fan exhibit high rock structure maturity with smaller particle sizes (0.120–0.040 mm versus 0.5–0.050 mm), high roundness and good sorting when compared with samples from the Ledong channel, indicating that the overall particle transport distance was longer [50,51].

In contrast, the Ledong sandstone is poorly sorted with a larger particle size than the Dongfang sandstone. Gravels in massive sandstone are often observed in the Ledong core sections. The Ledong rock structure maturity is low, reflecting features of near-source accumulation. The lengths (average 487 km) and drainage areas (average 80,000 km^2) of rivers from the northwestern provenance, such as the Red River and Lam River, are larger than those of the rivers of Hainan Island, and the distance required for transportation from the source to the sedimentation area was longer [36,52] (Figure 1a). Therefore, the source of the Dongfang region more likely originated from the northwestern provenance.

The rock composition study suggests that the Ledong provenance is more likely to have originated from Hainan Island, which is rich in magmatic rocks. Both the Ledong and Dongfang sandstone compositions are dominated by quartz, feldspar and rock debris (Figure 3). The Ledong rock debris is mainly composed of intrusive rocks. In contrast, the Dongfang rock debris contains mainly metamorphic rocks (average 9%), and the content of the argillaceous matrix is relatively low (1%), reflecting higher maturity of the rock composition and longer particle transport distance characteristics [50,53]. The most significant difference lies in the content of polycrystalline quartz. The content of polycrystalline quartz in the Ledong sample slices (20%) is higher than that in the Dongfang samples (6%). Researchers have found that polycrystalline quartz grains with similar sizes, sub-equiaxial morphologies and non-oriented arrangement characteristics are most likely derived from granite [54]. The polycrystalline quartz grains found within the thin Ledong sections have shown the above features and are mostly contacted by sutures, reflecting a granite origin. Compared with the Yangtze plate and the Indian Plate, Hainan Island's magmatic rocks are widely distributed (Figure 2), with an exposed area of approximately 17,000 km^2, covering 45% of the whole island [55]. Mesozoic–Paleozoic granites are mainly developed on the central and southern islands, and a large number of Cenozoic basalts appear in the north. These magmatic rocks were mainly formed during the Hercynian–Indosinian period, when volcanic magmatism was frequent and intense [45]. Therefore, the Ledong source is more likely to have Hainan provenance.

With respect to heavy mineral assemblage (Figure 4), the Dongfang well samples are more similar to the Lam, Ma, Giang and Red River samples, which are characterized by high leucoxene, low zircon and low ZTR values. The northwestern provenance presents relatively low maturity of heavy minerals (ZTR% 28.39), as fewer magmatic parent rocks are exposed in eastern Vietnam and the southern Yangtze plate (Figure 2), accounting for approximately 5% [41], so the zircon content from the northwestern provenance is lower than that from the Hainan provenance. The Ledong well samples are more similar to Hainan rivers, such as the Ningyuan, Wanglou, Changhua, Ganen and Beili Rivers, which are characterized by having high zircon, high magnetite, high garnet, high ZTR and low leucoxene values (Figure 4). The Hainan provenance has a high degree of maturity of heavy minerals (ZTR 41.46%). The abundant zircon and magnetite in the Hainan River sands come from the intermediately acidic granites exposed on the island [32].

Since the Miocene, the sediment compositions provided by the rivers around the Yinggehai Basin have not remained exactly constant but have been influenced by tectonics, weathering and other factors [2,56,57]. However, in this research, we sampled the modern river deposits to compare with the Miocene (ca. 8.2–5.3 Ma) sediments in the basin from the wells based on the assumption that the rivers around Yinggehai Basin had a similar detrital zircon signature in the Miocene as they do at present. This assumption is based on the following evidence. Since the late Miocene, no large-scale plate collision or river capture events have been recorded [7]. After the collision of the Indosinian–Eurasian plate and the uplift of the Qinghai-Tibet orogenic belt in the late Paleocene to early Eocene, the provenance characteristics of the Yangtze plate, Indosinian plate and South China plate have been relatively stable [40,58]. Previous research has found that the Red River was captured by the Yangtze River before the late early Miocene [37]. However, since the late Miocene, no such large-scale river change events have occurred in the surrounding basins of Yinggehai basin. In addition, many scholars have also used modern river samples to replace specific ancient river samples when studying the provenance of Yinggehai basin, which has been proved to be a reliable way [21,36]. Therefore, river sediments around the Yinggehai Basin have maintained relatively stable geochemical characteristics, and the age distribution of river sediments in the late Miocene is similar to that of modern river sediments.

The zircon U-Pb dating results are well matched with that of the lithological distribution in the source area. The lithologies exposed in the Yangtze block and the Indo-China block in the northwestern part of the basin are relatively complex, ranging from Cenozoic to Archaean strata [24,59] (Figure 2). Therefore, the Lam, Ma, Giang and Red Rivers are characterized by multiple high Cenozoic (33–37 Ma), Mesozoic (243–262 Ma) and Paleozoic (420–456 Ma) peaks and by multiple low Proterozoic peaks, such as 735–1028 Ma, 1700–1853 Ma and 2471–2490 Ma (Figure 5), mainly from Paleoproterozoic, Meso-Neoproterozoic deep metamorphic rocks and Meso-Neoproterozoic volcanic rocks from the northwestern provenance [44,60]. Cretaceous strata are widely distributed in the southwestern Hainan Island, and Permian to Triassic granites are largely exposed in the northwestern Hainan Island, with only a few Mesozoic to Paleozoic strata remaining in island blocks in the granite zone throughout the island [21]. Therefore, the Hainan provenance has two distinct peaks: Cretaceous and Permian–Triassic (Figure 5). The Cretaceous peaks are more prominent in the Ningyuan and Wanglou Rivers, which are located on the southwestern margin of Hainan Island, whereas the Permian–Triassic peaks are more prominent in the Changhua River on the northwestern margin of Hainan Island (Figure 5). It can be intuitively seen from the MDS diagram that the characteristics of the Hainan provenance and the northwestern provenance are clearly distinguished (Figure 8). The zircon age signatures of sandstone samples from Well LD-a and Well LD-b are closely linked to the Hainan provenance, whereas the Well DF-a and Well DF-b sandstone samples are adjacent to the northwestern provenance.

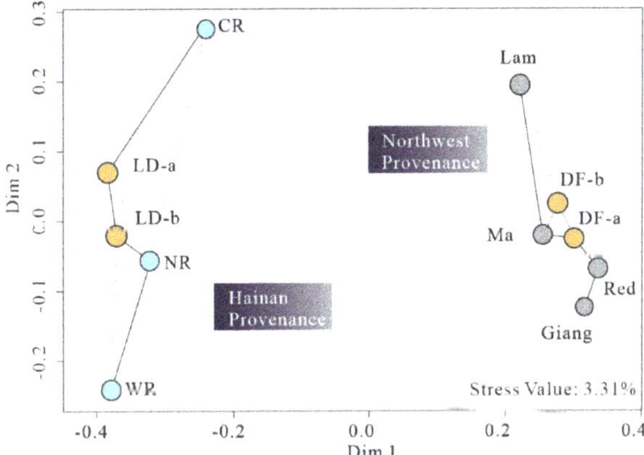

Figure 8. Multidimensional scaling (MDS) plot of detrital zircon U-Pb ages in river and study block sediments. The distance between samples was positively correlated with the Kolmogorov–Smirnov dissimilarity of their dating data. The solid lines connect the nearest neighbors, and the dotted lines link the second nearest samples. The stress value of 3.31% demonstrates an acceptable goodness of fit. The yellow, blue and gray dots represent the borehole samples, Hainan River samples and northwestern river samples, respectively.

5.2. Route to Source Comparison

A previous study found that the majority of continental-margin deep-water depositions occurred during periods of low sea levels [2]. The relative sea level decline was one of the most important factors triggering the development of marine gravity-driven systems in the basin during the Miocene [57,61]. The large-scale sea-level decline in the early depositional stage of the Huangliu Formation in the South China Sea has been confirmed by paleontological and geochemical evidence [56]. This feature is consistent with global sea-level changes (Figure 1b). In the early Miocene, when the sea level rose and accommodation space increased, deltas were extensively developed around the basin on the continental shelf due to continuous feeding from provenance rivers [38,60] (Figure 9a). The Yinggehai Basin strata experienced several relative sea level declines during the deposition process, among which the relative sea level declined sharply at the beginning of the late Miocene Huangliu Formation, and this decline was the largest in all geologic periods (Figure 1b), which was conducive to the transport of deltaic sediments along the land slope and the formation of gas reservoir deposits in the basin [22,62]. The Central Canyon in the Qiongdongnan Basin of the South China Sea and the second phase of the Bengal–Nicobar Fan in the Bay of Bengal were also initially developed ca. 10.5 Ma [1,63]. The thickness and connectivity of the sand body in the first stage are better than those in the second stage, during which the sea level gradually rose (Figure 7). Typical gravity flow deposition structures, such as the Bouma sequence, massive bedding and deformed bedding structures, can be observed (Figure 3) in both core sections. Multiple sets of thick sand body reservoirs are sandwiched between mud layers in both the Ledong and Dongfang wells, and the lithology fines upward (Figure 3), reflecting that the overall hydrodynamic conditions have changed from strong to weak [4,27].

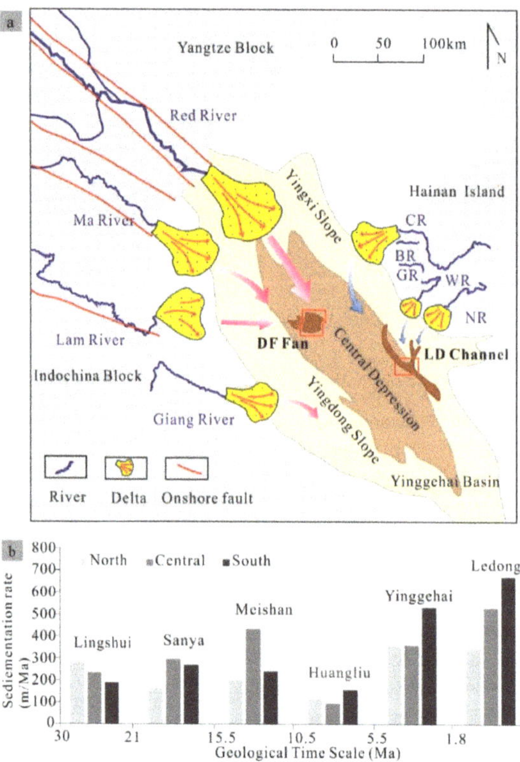

Figure 9. (a) Late Miocene sedimentary model for the Ledong submarine channel and Dongfang submarine fan in the Yinggehai Basin Huangliu Formation. The red arrow indicates northwestern provenance migration, whereas the blue arrow indicates Hainan provenance migration. NR: Ningyuan River, WR: Wanglou River, GR: Ganen River, BR: Beili River, CR: Changhua River. (b) Yinggehai Basin sedimentation rate in the northern, central and southern regions (data modified from Liu et al. (2017) [64]).

The source transport pathways of paleorivers are not very different from those of modern rivers. In the Miocene, the fault zone around the basin controlled the development of the paleotectonic "trench" and provided a path for the transportation of the provenance [64]. The Yingxi Fault zone, consisting of the Da River Fault, Lam River Fault, Ma River Fault and Troung Son Fault, strikes in the SSE–SE direction [30] (Figure 2). The Red River fault zone in the Yinggehai Basin strikes in the near-SN direction [60]. The paleotectonic "trenches" are well developed under the control of these two major fault zones. For example, the paleo Red River, Ma River and Lam River trenches are well superimposed (Figure 9a) with the Red River Fault, Ma River Fault and Lam River Fault, with an SSE–SE striking direction as well [64]. Therefore, the Yingxi Fault zone and the Red River Fault zone controlled the development and trends of ancient rivers. The ancient and modern Red River, Ma River and Lam River all developed along the SSE–SN fault zone [65,66].

Since ca. 30 Ma, the sedimentary center of the Yinggehai Basin migrated from northwest to southeast [26,38]. In the late Miocene Huanliu Formation, the sedimentation center was in the southern part of the basin, which prompted southeastward gravity flow deposition, as shown in Figure 9b. The migration is evidenced by the changes in sedimentation rate in the basin from the Oligocene to the present [67]. During the late Oligocene (30–21 Ma), the sedimentary center of the basin was located in the Lingao area in the north, with a deposition rate of 278 m/Ma (Figure 9b). The depositional center migrated to the central region in the early Miocene Sanya Formation to Meishan Formation deposition period (21–10.5 Ma), with increasing deposition rates of 294 m/Ma and 433 m/Ma, respectively. During the deposition of the Miocene Huangliu Formation, the deposition rate in the south gradually exceeded the deposition rate in the center, and the deposition center began to migrate further southward with a relatively low sedimentation rate of 155 m/Ma [67]. From the deposition period of the Pliocene Yinggehai Formation to the Ledong Formation (5.5 Ma—present), the deposition rate of the southern area increased rapidly, reaching 531 to 670 m/Ma.

The gravity flow systems developed in the study areas of Ledong and Dongfang are also greatly influenced by regional tectono-morphologic factors, such as regional faulting activities, in situ paleotopography and the style of slope break belts developed on the continental margin. During the depositional period of the Huangliu Formation, the paleo-geomorphic environment and the circumferential tectonic activity in the Ledong and Dongfang regions had a clear controlling role on the formation, evolution, development and spatial distribution characteristics of the sedimentary systems. The submarine channel developed in the eastern part of the basin is mainly affected by the faulted slope break zone controlled by the No. 1 Fault zone, and the submarine fan in the western part of the basin is affected by the flexure slope break zone controlled by the Yingxi Fault zone (Figure 10). The violent movement of the No. 1 Fault led to regional tectonic subsidence and the formation of a faulted slope break belt on the eastern margin of the basin [6,28]. The Yingdong Slope in the east of the basin is dominated by the No. 1 Fault, with a high level of faulting activity. The average activity rates of the southern section of the No. 1 Fault in the Sanya Formation (21–15.5 Ma), early Meishan Formation (15.5–13.4 Ma), late Meishan Formation (13.4–10.5 Ma), early Huangliu Formation (10.5–8.2 Ma) and late Huangliu Formation (8.2–5.5 Ma) are approximately 49, 57, 78, 35 and 46 m/Ma, respectively [68]. The faulted slope break belt has a high slope–fall ratio (Figure 10), and the fault-descent plate forms a certain sedimentary slope toward the center of the basin. Thus, the paleomorphological slope of the Ledong district is relatively steep, which easily triggers the development of gravity flow along the clinoform [69]. At the same time, hidden faults are traced under the T40 seismic interface at the bottom of the Ledong channel (Figure 6c), making the strata in this area more susceptible to the later transformation of terrigenous debris and currents to form a gravity flow channel. Therefore, the development of the northwestern trend tributary path is restricted by the low terrain created by the southeastern trend bottom faults, and the northeastern trend tributary sediment transportation route is along the faulted slope break. The Yingxi Fault zone in the western part of the basin is controlled by several medium- to small-scale normal faults (Figure 10), with average slopes ranging from 3° to 5°, among which the first-level flexural slope break is controlled by some medium faults, such as the Lam River fault, Ma River fault and Da River Fault [26]. This geomorphic feature makes the distribution of the sedimentary system to be mainly controlled by the rise and fall of sea level, the location of slope break points and the migration of the deposition center. During the late Miocene, the deposition center was in the southeastern part of the basin, and the paleogeomorphic slope in the Dongfang area tended to be gentle, leaning steadily toward the SSE direction. The SSE-trending Dongfang submarine fan sedimentary system developed in the open and flat area.

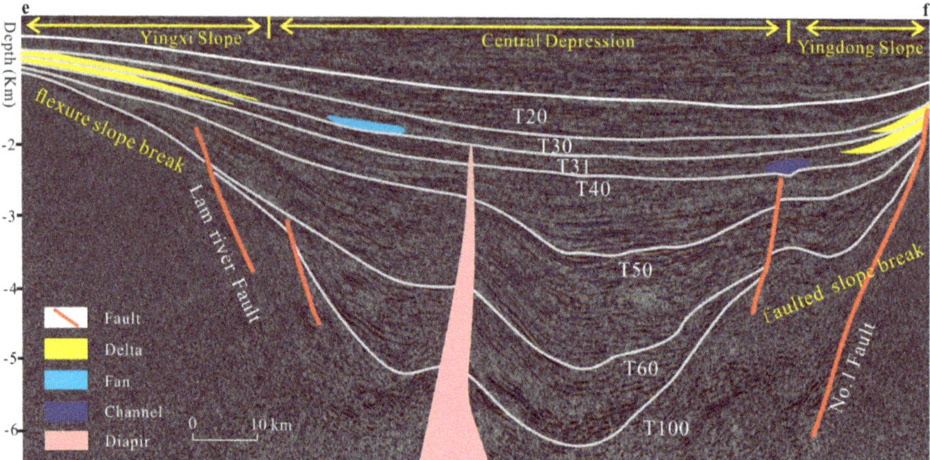

Figure 10. The flexure slope break zone was controlled by the Yingxi fault zone in the western margin of the basin, and the faulted slope break zone was controlled by the No. 1 Fault zone in the eastern margin of the basin (see the location of lines e–f in Figure 1).

6. Conclusions

From the perspective of source-to-sink, this study comparatively analyzed the provenance sources, formation mechanisms and sedimentary characteristics of the late Miocene axially restricted channel in the Ledong district and the submarine fan in the Dongfang district. The unique regional tectonic pattern of the Yinggehai Basin and the abundant source supplies are the dominant factors in the formation of the gravity flow source-to-sink sedimentary system and gas reservoirs in the study area. The sedimentary systems were generally influenced by the coupling control of sea level decline in the Miocene and the high-north low-south paleomorphology, which facilitated the debris migration into the basin and southward transport. The existence of a slope break zone in the margin of the basin promoted the formation of continuing gravity flows that were unloaded near the deposition center. In addition, the spatial variations within the basin controlled the deposition processes.

The source supply of the Ledong channel on the Yingdong Slope had Hainan provenance from the eastern part of the basin. With a short sediment transport distance, the rock structure is less mature, and the content of polycrystalline quartz in the rock composition is relatively high, showing a magmatic genesis. The heavy mineral assemblages are marked by high zircon, high ZTR and low leucoxene values. The zircon U-Pb age is characterized by the bimodal form of Cretaceous and Permian–Triassic peaks. The eastern margin slope break was affected by higher faulting intensity in the southern section of the No. 1 Fault and developed a faulted slope break zone characterized by relatively large and steep slopes, creating a favorable transport pathway for the rapid deposition of deltaic sands. The regional hidden fracture zone built up a natural low topography for the development of restricted channels.

The Dongfang submarine fan located in the Central Depression was fed by rivers in the northwestern margin of the basin, including the Red River and the rivers of eastern Vietnam. The Dongfang sediments were transported over long distances and present a higher maturity of rock structure than the Ledong sediments. The heavy minerals are characterized by high leucoxene, low zircon and low ZTR. Zircon age spectra show a multipeak coexistence. Affected by the Yingxi fault zone, the northwestern margin of the basin developed several normal faults and formed a gently sloping flexure slope break belt. When the sea level dropped, the debris was transported through the flexural slope break belt to the relatively flat and open area below the break point and was deposited to form a seafloor fan tilting to the SSE.

Author Contributions: Conceptualization, investigation and data curation, Y.Y. and Q.G.; methodology, Q.G. and H.W.; formal analysis, validation, software, resources and visualization, Y.Y.; writing—original draft preparation, Y.Y. and Q.G.; writing—review and editing, Y.Y. and H.W.; supervision, H.W.; project administration and funding acquisition, H.W. and Q.G. All authors have read and agreed to the published version of the manuscript.

Funding: This research was funded by the Key Program of the National Natural Science Foundation of China (grant numbers U19B2007) and the scholarship of China Postdoctoral Foundation (grant numbers 2021M700201).

Institutional Review Board Statement: Not applicable.

Informed Consent Statement: Not applicable.

Data Availability Statement: Not applicable.

Acknowledgments: The authors would like to acknowledge the China National Offshore Oil Corporation (Haikou and Zhanjiang Branch).

Conflicts of Interest: The authors declare no conflict of interest.

References

1. Gong, C.; Steel, R.J.; Qi, K.; Wang, Y. Deep-Water Channel Morphologies, Architectures, and Population Densities in Relation to Stacking Trajectories and Climate States. *Bull. Geol. Soc. Am.* **2020**, *133*, 287–306. [CrossRef]
2. Covault, J.A.; Graham, S.A. Submarine Fans at All Sea-Level Stands: Tectono-Morphologic and Climatic Controls on Terrigenous Sediment Delivery to the Deep Sea. *Geology* **2010**, *38*, 939–942. [CrossRef]
3. Shanmugam, G. Deep-Marine Tidal Bottom Currents and Their Reworked Sands in Modern and Ancient Submarine Canyons. *Mar. Pet. Geol.* **2003**, *20*, 471–491. [CrossRef]
4. Shanmugam, G. *Submarine Fans: A Critical Retrospective (1950–2015)*; Elsevier Ltd.: Amsterdam, The Netherlands, 2016; Volume 5, ISBN 3695110139.
5. Yu, K.; Miramontes, E.; Alves, T.M.; Li, W.; Liang, L.; Li, S.; Zhan, W.; Wu, S. Incision of Submarine Channels Over Pockmark Trains in the South China Sea. *Geophys. Res. Lett.* **2021**, *48*, e2021GL092861. [CrossRef]
6. Sun, M.; Wang, H.; Liao, J.; Gan, H.; Xiao, J.; Ren, J.; Zhao, S. Sedimentary Characteristics and Model of Gravity Flow Depositional System for the First Member of Upper Miocene Huangliu Formation in Dongfang Area, Yinggehai Basin, Northwestern South China Sea. *J. Earth Sci.* **2014**, *25*, 506–518. [CrossRef]
7. Wang, C.; Liang, X.; Xie, Y.; Tong, C.; Pei, J.; Zhou, Y.; Jiang, Y.; Fu, J.; Dong, C.; Liu, P. Provenance of Upper Miocene to Quaternary Sediments in the Yinggehai-Song Hong Basin, South China Sea: Evidence from Detrital Zircon U-Pb Ages. *Mar. Geol.* **2014**, *355*, 202–217. [CrossRef]
8. Gong, C.; Li, D.; Steel, R.J.; Peng, Y.; Xu, S.; Wang, Y. Delta-to-Fan Source-to-Sink Coupling as a Fundamental Control on the Delivery of Coarse Clastics to Deepwater: Insights from Stratigraphic Forward Modelling. *Basin Res.* **2021**, *33*, 2960–2983. [CrossRef]
9. Cathro, D.L.; Austin, J.A.; Moss, G.D. Progradation along a Deeply Submerged Oligocene-Miocene Heterozoan Carbonate Shelf: How Sensitive Are Clinoforms to Sea Level Variations? *Am. Assoc. Pet. Geol. Bull.* **2003**, *87*, 1547–1574. [CrossRef]
10. Fan, C.; Cao, J.; Luo, J.; Li, S.; Wu, S.; Dai, L.; Hou, J.; Mao, Q. Heterogeneity and Influencing Factors of Marine Gravity Flow Tight Sandstone under Abnormally High Pressure: A Case Study from the Miocene Huangliu Formation Reservoirs in LD10 Area, Yinggehai Basin, South China Sea. *Pet. Explor. Dev.* **2021**, *48*, 1048–1062. [CrossRef]
11. Duan, W.; Li, C.-F.; Chen, X.-G.; Luo, C.-F.; Tuo, L.; Liu, J.-Z. Diagenetic Differences Caused by Gas Charging with Different Compositions in the XF13 Block of the Yinggehai Basin, South China Sea. *AAPG Bull.* **2020**, *104*, 735–765. [CrossRef]

12. Huang, Y.; Tan, X.; Liu, E.; Wang, J.; Wang, J. Sedimentary Processes of Shallow-Marine Turbidite Fans: An Example from the Huangliu Formation in the Yinggehai Basin, South China Sea. *Mar. Pet. Geol.* **2021**, *132*, 105191. [CrossRef]
13. Huang, Y.; Kane, I.A.; Zhao, Y. Effects of Sedimentary Processes and Diagenesis on Reservoir Quality of Submarine Lobes of the Huangliu Formation in the Yinggehai Basin, China. *Mar. Pet. Geol.* **2020**, *120*, 104526. [CrossRef]
14. Craddock, W.H.; Kylander, A.R.C. U-Pb Ages of Detrital Zircons from the Tertiary Mississippi River Delta in Central Louisiana: Insights into Sediment Provenance. *Geosphere* **2013**, *9*, 1832–1851. [CrossRef]
15. Cerri, R.I.; Warren, L.V.; Varejão, F.G.; Marconato, A.; Luvizotto, G.L.; Assine, M.L. Unraveling the Origin of the Parnaíba Basin: Testing the Rift to Sag Hypothesis Using a Multi-Proxy Provenance Analysis. *J. S. Am. Earth Sci.* **2020**, *101*, 102625. [CrossRef]
16. Zhao, L.; Liu, X.; Yang, S.; Ma, X.; Liu, L.; Sun, X. Regional Multi-Sources of Carboniferous Karstic Bauxite Deposits in North China Craton: Insights from Multi-Proxy Provenance Systems. *Sediment. Geol.* **2021**, *421*, 105958. [CrossRef]
17. Caracciolo, L. Sediment Generation and Sediment Routing Systems from a Quantitative Provenance Analysis Perspective: Review, Application and Future Development. *Earth-Sci. Rev.* **2020**, *209*, 103226. [CrossRef]
18. Allen, P.A. From Landscapes into Geological History. *Nature* **2008**, *451*, 274–276. [CrossRef]
19. Liu, Z.; Zhao, Y.; Colin, C.; Stattegger, K.; Wiesner, M.G.; Huh, C.A.; Zhang, Y.; Li, X.; Sompongchaiyakul, P.; You, C.F.; et al. Source-to-Sink Transport Processes of Fluvial Sediments in the South China Sea. *Earth-Sci. Rev.* **2016**, *153*, 238–273. [CrossRef]
20. Walsh, J.P.; Wiberg, P.L.; Aalto, R.; Nittrouer, C.A.; Kuehl, S.A. Source-to-Sink Research: Economy of the Earth's Surface and Its Strata. *Earth-Sci. Rev.* **2016**, *153*, 1–6. [CrossRef]
21. Cao, L.; Jiang, T.; Wang, Z.; Zhang, Y.; Sun, H. Provenance of Upper Miocene Sediments in the Yinggehai and Qiongdongnan Basins, Northwestern South China Sea: Evidence from REE, Heavy Minerals and Zircon U-Pb Ages. *Mar. Geol.* **2015**, *361*, 136–146. [CrossRef]
22. Jiang, T.; Cao, L.; Xie, X.; Wang, Z.; Li, X.; Zhang, Y.; Zhang, D.; Sun, H. Insights from Heavy Minerals and Zircon U-Pb Ages into the Middle Miocene-Pliocene Provenance Evolution of the Yinggehai Basin, Northwestern South China Sea. *Sediment. Geol.* **2015**, *327*, 32–42. [CrossRef]
23. Yao, Y.; Guo, S.; Zhu, H.; Huang, Y.; Liu, H.; Wang, X. Source-to-Sink Characteristics of the Channelized Submarine Fan System of the Huangliu Formation in the Dongfang Block, Yinggehai Basin, South China Sea. *J. Pet. Sci. Eng.* **2021**, *206*, 109009. [CrossRef]
24. Clift, P.D.; Sun, Z. The Sedimentary and Tectonic Evolution of the Yinggehai-Song Hong Basin and the Southern Hainan Margin, South China Sea: Implications for Tibetan Uplift and Monsoon Intensification. *J. Geophys. Res. Solid Earth* **2006**, *111*, 1–28. [CrossRef]
25. Li, X.; Wang, H.; Yao, G.; Peng, Y.; Ouyang, J.; Zhao, X.; Lan, Z.; Huang, C. Sedimentary Features and Filling Process of the Miocene Gravity-Driven Deposits in Ledong Area, Yinggehai Basin, South China Sea. *J. Pet. Sci. Eng.* **2021**, *209*, 109886. [CrossRef]
26. Wang, Z.; Xian, B.; Liu, J.; Fan, C.; Li, H.; Wang, J.; Zhang, X.; Huang, H.; Tan, J.; Chen, P.; et al. Large-Scale Turbidite Systems of a Semi-Enclosed Shelf Sea: The Upper Miocene of Eastern Yinggehai Basin, South China Sea. *Sediment. Geol.* **2021**, *425*, 106006. [CrossRef]
27. Huang, Y.; Yao, G.; Fan, X. Sedimentary Characteristics of Shallow-Marine Fans of the Huangliu Formation in the Yinggehai Basin, China. *Mar. Pet. Geol.* **2019**, *110*, 403–419. [CrossRef]
28. Xie, X.; Müller, R.D.; Ren, J.; Jiang, T.; Zhang, C. Stratigraphic Architecture and Evolution of the Continental Slope System in Offshore Hainan, Northern South China Sea. *Mar. Geol.* **2008**, *247*, 129–144. [CrossRef]
29. Ding, W.; Hou, D.; Zhang, W.; He, D.; Cheng, X. A New Genetic Type of Natural Gases and Origin Analysis in Northern Songnan-Baodao Sag, Qiongdongnan Basin, South China Sea. *J. Nat. Gas Sci. Eng.* **2018**, *50*, 384–398. [CrossRef]
30. Lei, C.; Ren, J.; Sternai, P.; Fox, M.; Willett, S.; Xie, X.; Clift, P.D.; Liao, J.; Wang, Z. Structure and Sediment Budget of Yinggehai-Song Hong Basin, South China Sea: Implications for Cenozoic Tectonics and River Basin Reorganization in Southeast Asia. *Tectonophysics* **2015**, *655*, 177–190. [CrossRef]
31. Qin, Y.; Alves, T.M.; Constantine, J.A.; Gamboa, D.; Wu, S. Effect of Channel Tributaries on the Evolution of Submarine Channel Confluences (Espírito Santo Basin, SE Brazil). *Bull. Geol. Soc. Am.* **2020**, *132*, 263–272. [CrossRef]
32. Gong, Y.; Pease, V.; Wang, H.; Gan, H.; Liu, E.; Ma, Q.; Zhao, S.; He, J. Insights into Evolution of a Rift Basin: Provenance of the Middle Eocene-Lower Oligocene Strata of the Beibuwan Basin, South China Sea from Detrital Zircon. *Sediment. Geol.* **2021**, *419*, 105908. [CrossRef]
33. He, J.; Garzanti, E.; Cao, L.; Wang, H. The Zircon Story of the Pearl River (China) from Cretaceous to Present. *Earth-Sci. Rev.* **2020**, *201*, 103078. [CrossRef]
34. Yang, J.; Huang, B. Origin and Migration Model of Natural Gas in L Gas Field, Eastern Slope of Yinggehai Sag, China. *Pet. Explor. Dev.* **2019**, *46*, 471–481. [CrossRef]
35. Fan, C. Tectonic Deformation Features and Petroleum Geological Significance in Yinggehai Large Strike-Slip Basin, South China Sea. *Pet. Explor. Dev.* **2018**, *45*, 190–199. [CrossRef]
36. Wang, C.; Liang, X.; Foster, D.A.; Xie, Y.; Tong, C.; Pei, J.; Fu, J.; Jiang, Y.; Dong, C.; Zhou, Y.; et al. Zircon U-Pb Geochronology and Heavy Mineral Composition Constraints on the Provenance of the Middle Miocene Deep-Water Reservoir Sedimentary Rocks in the Yinggehai-Song Hong Basin, South China Sea. *Mar. Pet. Geol.* **2016**, *77*, 819–834. [CrossRef]
37. Wang, C.; Liang, X.; Foster, D.A.; Tong, C.; Liu, P.; Liang, X.; Zhang, L. Linking Source and Sink: Detrital Zircon Provenance Record of Drainage Systems in Vietnam and the Yinggehai-Song Hong Basin, South China Sea. *Bull. Geol. Soc. Am.* **2019**, *131*, 191–204. [CrossRef]

38. Wang, C.; Liang, X.; Foster, D.A.; Liang, X.; Tong, C.; Liu, P. Detrital Zircon Ages: A Key to Unraveling Provenance Variations in the Eastern Yinggehai-Song Hong Basin, South China Sea. *Am. Assoc. Pet. Geol. Bull.* **2019**, *103*, 1525–1532. [CrossRef]
39. Chen, Y.; Zhang, D.; Zhang, J.; Huang, C.; Jiao, X.; Wang, Y. Sedimentary Characteristics and Controlling Factors of the Axial Gravity Channel in Huangliu Formation of the Yingdong Slope Area in Yinggehai Basin. *J. Northeast Pet. Univ.* **2020**, *44*, 91–102.
40. Yan, Y.; Carter, A.; Palk, C.; Brichau, S.; Hu, X. Understanding Sedimentation in the Song Hong-Yinggehai Basin, South China Sea. *Geochem. Geophys. Geosyst.* **2011**, *12*. [CrossRef]
41. Nguyen, H.H.; Carter, A.; Van Hoang, L.; Vu, S.T. Provenance, Routing and Weathering History of Heavy Minerals from Coastal Placer Deposits of Southern Vietnam. *Sediment. Geol.* **2018**, *373*, 228–238. [CrossRef]
42. Xie, Y.; Li, X.; Fan, C.; Tan, J.; Liu, K.; Lu, Y.; Hu, W.; Li, H.; Wu, J. The Axial Channel Provenance System and Natural Gas Accumulation of the Upper Miocene Huangliu Formation in Qiongdongnan Basin, South China Sea. *Pet. Explor. Dev.* **2016**, *43*, 570–578. [CrossRef]
43. Zhong, L.; Li, G.; Yan, W.; Xia, B.; Feng, Y.; Miao, L.; Zhao, J. Using Zircon U–Pb Ages to Constrain the Provenance and Transport of Heavy Minerals within the Northwestern Shelf of the South China Sea. *J. Asian Earth Sci.* **2017**, *134*, 176–190. [CrossRef]
44. Fyhn, M.B.W.; Thomsen, T.B.; Keulen, N.; Knudsen, C.; Rizzi, M.; Bojesen-Koefoed, J.; Olivarius, M.; Tri, T.V.; Phach, P.V.; Minh, N.Q.; et al. Detrital Zircon Ages and Heavy Mineral Composition along the Gulf of Tonkin—Implication for Sand Provenance in the Yinggehai-Song Hong and Qiongdongnan Basins. *Mar. Pet. Geol.* **2019**, *101*, 162–179. [CrossRef]
45. Pe-Piper, G.; Piper, D.J.W.; Wang, Y.; Zhang, Y.; Trottier, C.; Ge, C.; Yin, Y. Quaternary Evolution of the Rivers of Northeast Hainan Island, China: Tracking the History of Avulsion from Mineralogy and Geochemistry of River and Delta Sands. *Sediment. Geol.* **2016**, *333*, 84–99. [CrossRef]
46. Jiang, T.; Liu, X.; Yu, T.; Hu, Y. OSL Dating of Late Holocene Coastal Sediments and Its Implication for Sea-Level Eustacy in Hainan Island, Southern China. *Quat. Int.* **2018**, *468*, 24–32. [CrossRef]
47. Huang, B.; Xiao, X.; Hu, Z.; Yi, P. Geochemistry and Episodic Accumulation of Natural Gases from the Ledong Gas Field in the Yinggehai Basin, Offshore South China Sea. *Org. Geochem.* **2005**, *36*, 1689–1702. [CrossRef]
48. Andò, S.; Garzanti, E.; Padoan, M.; Limonta, M. Corrosion of Heavy Minerals during Weathering and Diagenesis: A Catalog for Optical Analysis. *Sediment. Geol.* **2012**, *280*, 165–178. [CrossRef]
49. Hubert, J.F. A Zircon-Tourmaline-Rutile Maturity Index and the Interdependence of the Composition of Heavy Mineral Assemblages with the Gross Composition and Texture of Sandstones. *SEPM J. Sediment. Res.* **1962**, *32*, 440–450. [CrossRef]
50. Garzanti, E. Petrographic Classification of Sand and Sandstone. *Earth-Sci. Rev.* **2019**, *192*, 545–563. [CrossRef]
51. Garzanti, E. From Static to Dynamic Provenance Analysis-Sedimentary Petrology Upgraded. *Sediment. Geol.* **2016**, *336*, 3–13. [CrossRef]
52. Milliman, J.D.; Farnsworth, K.L. *River Discharge to the Coastal Ocean: A Global Synthesis*; Cambridge University Press: Cambridge, UK, 2011; ISBN 9780511781247.
53. Baas, J.H.; Hailwood, E.A.; McCaffrey, W.D.; Kay, M.; Jones, R. Directional Petrological Characterisation of Deep-Marine Sandstones Using Grain Fabric and Permeability Anisotropy: Methodologies, Theory, Application and Suggestions for Integration. *Earth-Sci. Rev.* **2007**, *82*, 101–142. [CrossRef]
54. Wu, Y.; Zheng, Y. Genesis of Zircon and Its Constraints on Interpretation of U-Pb Age. *Chin. Sci. Bull.* **2004**, *49*, 1554–1569. [CrossRef]
55. Hu, B.; Li, J.; Cui, R.; Wei, H.; Zhao, J.; Li, G.; Fang, X.; Ding, X.; Zou, L.; Bai, F. Clay Mineralogy of the Riverine Sediments of Hainan Island, South China Sea: Implications for Weathering and Provenance. *J. Asian Earth Sci.* **2014**, *96*, 84–92. [CrossRef]
56. Zhong, Y.; Wilson, D.J.; Liu, J.; Wan, S.; Bao, R.; Liu, J.; Zhang, Y.; Wang, X.; Liu, Y.; Liu, X.; et al. Contrasting Sensitivity of Weathering Proxies to Quaternary Climate and Sea-Level Fluctuations on the Southern Slope of the South China Sea. *Geophys. Res. Lett.* **2021**, *48*, e2021GL096433. [CrossRef]
57. Gong, C.; Blum, M.D.; Wang, Y.; Lin, C.; Xu, Q. Can Climatic Signals Be Discerned in a Deep-Water Sink?: An Answer from the Pearl River Source-to-Sink Sediment-Routing System. *Bull. Geol. Soc. Am.* **2018**, *130*, 661–677. [CrossRef]
58. Wissink, G.K.; Hoke, G.D.; Garzione, C.N.; Liu-Zeng, J. Temporal and Spatial Patterns of Sediment Routing across the Southeast Margin of the Tibetan Plateau: Insights from Detrital Zircon. *Tectonics* **2016**, *35*, 2538–2563. [CrossRef]
59. Van Hoang, L.; Wu, F.Y.; Clift, P.D.; Wysocka, A.; Swierczewska, A. Evaluating the Evolution of the Red River System Based on in Situ U-Pb Dating and Hf Isotope Analysis of Zircons. *Geochem. Geophys. Geosyst.* **2009**, *10*. [CrossRef]
60. Clift, P.D.; Van Long, H.; Hinton, R.; Ellam, R.M.; Hannigan, R.; Tan, M.T.; Blusztajn, J.; Duc, N.A. Evolving East Asian River Systems Reconstructed by Trace Element and Pb and Nd Isotope Variations in Modern and Ancient Red River-Song Hong Sediments. *Geochem. Geophys. Geosyst.* **2008**, *9*. [CrossRef]
61. Li, C.; Lyu, C.; Chen, G.; Zhang, G.; Ma, M.; Yang, H.; Bi, G. Zircon U-Pb Ages and REE Composition Constraints on the Provenance of the Continental Slope-Parallel Submarine Fan, Western Qiongdongnan Basin, Northern Margin of the South China Sea. *Mar. Pet. Geol.* **2019**, *102*, 350–362. [CrossRef]
62. Wang, H.; Chen, S.; Gan, H.; Liao, J.; Sun, M. Accumulation Mechanism of Large Shallow Marine Turbidite Deposits: A Case Study of Gravity Flow Deposits of the Huangliu Formation in Yinggehai Basin. *Earth Sci. Front.* **2015**, *22*, 21–34. [CrossRef]
63. Li, C.; Lv, C.; Chen, G.; Zhang, G.; Ma, M.; Shen, H.; Zhao, Z.; Guo, S. Source and Sink Characteristics of the Continental Slope-Parallel Central Canyon in the Qiongdongnan Basin on the Northern Margin of the South China Sea. *J. Asian Earth Sci.* **2017**, *134*, 1–12. [CrossRef]

64. Rangin, C.; Klein, M.; Roques, D.; Le Pichon, X.; Le Van, T. The Red River Fault System in the Tonkin Gulf, Vietnam. *Tectonophysics* **1995**, *243*, 209–222. [CrossRef]
65. Clift, P.; Lee, G.H.; Anh Duc, N.; Barckhausen, U.; Van Long, H.; Zhen, S. Seismic Reflection Evidence for a Dangerous Grounds Miniplate: No Extrusion Origin for the South China Sea. *Tectonics* **2008**, *27*, 1–16. [CrossRef]
66. Liang, L. A Study of the Genesis of Hainan Island. *Geol. China* **2018**, *45*, 693–705. [CrossRef]
67. Liu, W.; Pei, J.; Yu, J.; Fan, C.; Zhang, X.; Shao, Y. Palaeogeomorphologic Control on Gravity Flow Deposit Sand Gas Reservoir Formation: A Case Study from the Miocene in Yinggehai Basin. *Mar. Geol. Quat. Geol.* **2017**, *37*, 197–203.
68. Zhang, J.; Fan, C.; Tan, J.; Chen, Y.; Huang, C.; Luo, W. Evolution Characteristics of Sedimentary System in Yinggehai Basin in Miocene and Its Exploration Significance. *Geol. Sci. Technol. Inf.* **2019**, *38*, 51–59.
69. Pohl, F.; Eggenhuisen, J.T.; Cartigny, M.J.B.; Tilston, M.C.; de Leeuw, J.; Hermidas, N. The Influence of a Slope Break on Turbidite Deposits: An Experimental Investigation. *Mar. Geol.* **2020**, *424*, 106160. [CrossRef]

 *energies*

Article

Applicability Analysis of Pre-Stack Inversion in Carbonate Karst Reservoir

Rui Wang [1] **and Bo Liu** [2,*]

[1] Institute of Exploration and Development, SINOPEC Shanghai Offshore Oil & Gas Company, No. 1225, Mall Road, Pudong New District, Shanghai 200120, China; greatwr@126.com

[2] School of Earth Science and Engineering, Hebei University of Engineering, No. 19 Tai Chi Road, Economic and Technological Development Zone, Handan 056083, China

* Correspondence: liubo@hebeu.edu.cn

Abstract: Although pre-stack inversion has been carried out on reservoir prediction, few studies have focused on the application of pre-stack for seismic inversion in fractured-cavity carbonate reservoirs. In carbonate rock, complicated combinations and fluid predictions in karst caves are remain unclear. Post-stack methods are commonly used to predict the position, size, and fillings of caves, but pre-stack inversion is seldom applied in carbonate karst reservoirs. This paper proposes a pre-stack inversion method for forward modeling data and oil survey seismic data, using both points to indicate the application of pre-stack inversion in karst caves. Considering influence of cave size, depth, and filler on prediction, three sets of models (different caves volume; different fillings velocity of caves; complicated combination of caves) are employed and inverted by pre-stack inversion. We analyze the pre-stack results to depict Ordovician oil bearing and characterize caves. Geological model parameters came from actual data of the Tahe oilfield, and seismic data were synthesized from geological models based on full-wave equation forward simulation. Moreover, a case study of pre-stack inversion from the Tahe area was employed. The study shows that, from both the forward modeling and the oil seismic data points of view, pre-stack inversion is applicable to carbonate karst reservoirs.

Keywords: carbonate cavern reservoir application; pre-stack inversion; model

Citation: Wang, R.; Liu, B. Applicability Analysis of Pre-Stack Inversion in Carbonate Karst Reservoir. *Energies* **2022**, *15*, 5598. https://doi.org/10.3390/en15155598

Academic Editors: Yong Li, Fan Cui and Chao Xu

Received: 5 July 2022
Accepted: 27 July 2022
Published: 2 August 2022

Publisher's Note: MDPI stays neutral with regard to jurisdictional claims in published maps and institutional affiliations.

1. Introduction

Unconventional energy sources and those of deep oil and gas are showing increasing significance in energy structures, and the demand for exploration is increasing in China [1,2]. The Tarim Basin is a famous oil-rich basin in China with significant deep carbonate reserves, and is an important energy source for China. However, The Ordovician carbonate rocks in the target layer are buried at a depth of about 5500–6500 m, and fractured vuggy reservoirs show "string of beads" characteristics in the seismic section. The seismic responses of caves usually consist of one–three seismic peaks and troughs, which look like strings of beads. Therefore, it is called the beaded response [3–6]. Karst caves are generally made up by soft, porous fillings and fluids, surrounded by tight carbonate rocks. This composition leads to considerable impedance differences. In addition, the caves are limited not only vertically but also horizontally; as a result, their bright spots are limited both vertically and horizontally [7,8]. Caves are a main type of reservoir in the Tarim Basin [9]. Geologically, it is because of tectonic action that large faults were initially formed, and after uplifting and exposing the surface, they were leached by atmospheric water and eroded along the faults, thus forming favorable storage spaces. At present, the structure and distribution of fractured cavity reservoirs are not clear. Petroleum engineers must characterize combination and fluid-bearing caves and design wells to improve enhanced oil recovery (EOR) [10]. In recent years, due to the complexity of fractured cavity carbonate reservoirs and the multiresolution nature of seismic data, it is necessary to understand the prediction ability of seismic data, analyze the fillings or fluids in karst cavity reservoirs, and the improve quantitative analysis approaches for carbonate reservoirs.

Previous studies have shown that post-stack inversion has been commonly used for characterization of the top of caves, which has achieved certain results; however, research concerning the application of P-impedance elastic parameters to characterize caves is limited, and there are not many applications of other elastic parameters [11,12]. Previous researchers indicated that AVO (amplitude variation with offset) attribute analysis technology can be used to identify fluids in carbonate reservoirs, and the fluid prediction accuracy reached 83.3%, indicating that the elastic parameters information of pre-stack seismic data is more abundant, and it is a requirement for carrying out pre-stack seismic inversion research. Therefore, for the purposes of clarifying the quantitative prediction ability and reliability of pre-stack seismic inversion, this study addresses its use in a carbonate karst reservoir, in the interest of hydrocarbon exploration and development.

Pre-stack elastic impedance inversion contains more lithological and elastic impedance and physical information than post-stack impedance, and is obtained independently from pre-stack seismic data, rather than from acoustic impedance, as is the case in traditional post-stack inversion [13]. Thus, this makes up for the shortcomings of the acoustic impedance inversion results overlapping in the reservoir and increases the capacity of seismic inversion to predict the characteristics of reservoirs. Pre-stack inversion has rarely been applied to the Tarim Basin. A previous researcher [14] physically modeled the seismic response of a karst cave, and drew some conclusions surrounding the effects of the cave scale, velocity, spatial distribution, shape, and fluids on the "beads" and the corresponding relationships between six types of "beads" and karst caves; However, there is no validity analysis for pre-stack inversion. The authors of [15–19] used pre-stack inversion based on seismic data for an Ordovician TZ45 area, and indicated that pre-stack inversion can be applied for cave carbonate reservoir prediction. We note that previous researchers have not made a complete argument for the validity of pre-stack inversion in carbonate reservoirs.

In this study, we use pre-stack inversion from two perspectives to a carbonate cavern reservoir: the accuracy of the pre-stack inversion based on forward seismic data; application of pre-stack inversion based on measured seismic data in Tarim. First, three sets of geological models that take into consideration different cave sizes, cave buried depths, and fills in cave are constructed using the Tarim karst cave reservoir parameters Subsequently, Seismic inversion study by Jason Workbench was performed using synthetic seismic data that was generated by forward modeling using the full-wave equation on the basis of the above geological model [20–23]. Pre-stack inversion's characteristics are analyzed, along with the application impacts of pre- and post-stack inversion in complicated combination geological models and various fluid geological models [24,25]. Finally, the cave carbonate reservoirs in the Tahe area are predicted using pre-stack inversion. According to the findings of our study, pre-stack inversion is a viable method for predicting complicated fracture-cavity carbonate reservoirs since it significantly improves reservoir prediction accuracy.

2. Geological Modeling Design

The geological model in the paper was designed to simulate carbonate karst reservoirs with various velocities, depths, and scales. In this section, we describe the construction of our geological model in detail.

2.1. Model Design

The model design was based on both easy and difficult concepts and considers three cases: the external shape of the cave, the internal filling of the cave, and the cave system. According to the computing conditions and computing power required by the three-dimensional wave equation, the length of the smallest cave we designed is 10 m, and the lengths of other caves are different multiples of 10 m. The main frequency of the target layer is 30 Hz. The wavelength of a seismic wave in a cave would be 100 m based on a velocity of 3000 m/s, and the wavelength of a seismic wave in a cave would be 133 m based on a velocity of 4000 m/s; if the average speed of the upper layer of an Ordovician system is 4000–4500 m/s, then the wavelength of the earthquake wave in the Ordovician upper layer is 133–150 m. Therefore, when studying the resolution of a cave, the size of the cave can be set with a multiplied wavelength. Considering that the analysis needs to minimize the influence of structural relief on a given reservoir prediction, the stratigraphy frame tends to be designed with horizontal layers. The vertical depth of the entire three-dimensional model was controlled within 4 km.

In the Tarim Basin, the velocity of 4000 m/s is the average value of oil-bearing layers in carbonate reservoirs. In addition to reservoir filling fluid differences, the size of the caves also affects the accuracy of reservoir predictions. Therefore, we divided karst caves into three different sets by considering the complicated combination of characteristics and the accuracy of reservoir prediction. Table 1 gives the P wave velocity, density, and S wave velocity for 3 sets of models and 18 caves.

Table 1. Parameters of modeled caves.

Model No.	Cave No.	P Wave Velocity (m/s)	S Wave Velocity (m/s)	Density (kg/m^3)	Size/Distance (m)
1	1	2800	1700	2120	40×40
1	2	3150	2050	2250	40×40
1	3	3550	2600	2320	40×40
1	4	4200	2680	2400	40×40
1	5	4550	2700	2480	40×40
1	6	5150	3000	2560	40×40
2	7	4000	2670	2380	80×80
2	8	4000	2670	2380	40×40
2	9	4000	2670	2380	20×20
2	10	4000	2670	2380	10×10
3	11	4000	2670	2380	$160 \times 10/10$
3	12	4000	2670	2380	$160 \times 10/20$
3	13	4000	2670	2380	$160 \times 20/20$
3	14	4000	2670	2380	$160 \times 80/120$
3	15	4000	2670	2380	$160 \times 160/120$
3	16	4000	2670	2380	$80 \times 80/40$
3	17	4000	2670	2380	$20 \times 20/10$
3	18	4000	2670	2380	10×10

Geological model 1: Different fillings in the karst caves. Six caves were designed in model 1. Differences in seismic response caused by different fillings in karst caves. Figure 1a,d show depth domain geological model, Figure 1a shows a section of the model, Figure 1d shows a map of the model. The study area of model 1 is within the black dotted line. The six different colored caves in the model are the same size, and the volumes are all 40 m × 40 m × 40 m. Their filling velocities are 2800 m/s, 3150 m/s, 3550 m/s, 4200 m/s, 4550 m/s, and 5150 m/s; these are equivalent to gas, live oil, oil, oil–water, argillaceous, and calcareous mudstone fillings, respectively. More than 100 m separate the caves, ensuring that their seismic reflections are independent of one another.

Geological model 2: Different sizes and depths. Differing vertical and horizontal seismic resolutions were modeled for the caves. Figure 2a,d present a view of the depth domain and the models map: Figure 2a shows a section of the model; Figure 2d shows a map of the model. A series of models with varying volumes are designed to analyze the reflection characteristics. We based our study on the background of karst paleogeomorphology and previous research. The length, width, and height of the caves are 10 m, 20 m, 40 m, and 80 m, respectively, which correspond to lengths of 1/16, 1/8, 1/4, and 1/2 the wavelengths, in which the filling velocity is 4000 m/s. The depths from the top interface are between 0 m and 70 m. The distances of the caves from each other more than 100 m, which ensures that the seismic reflections between the caves do not interact with each other.

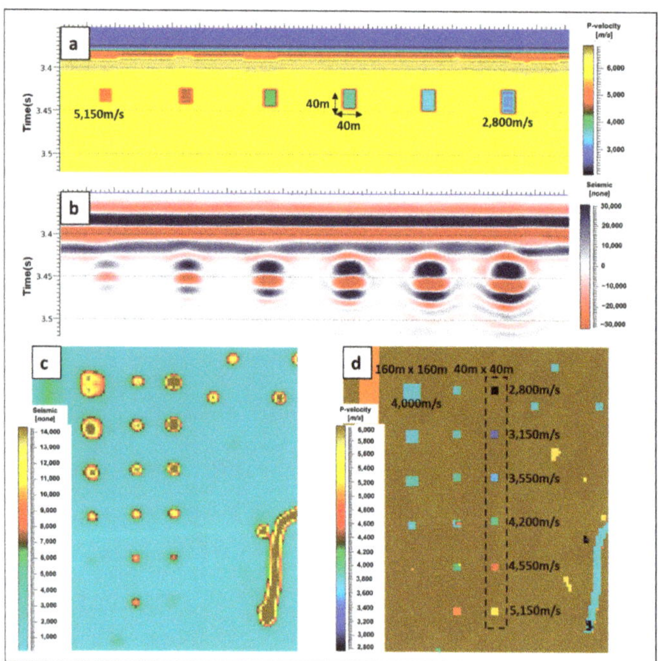

Figure 1. Models of different fillings in the caves. Geological model 1: (**a**,**d**) geological models; (**b**) seismic model; (**c**) RMS amplitude map.

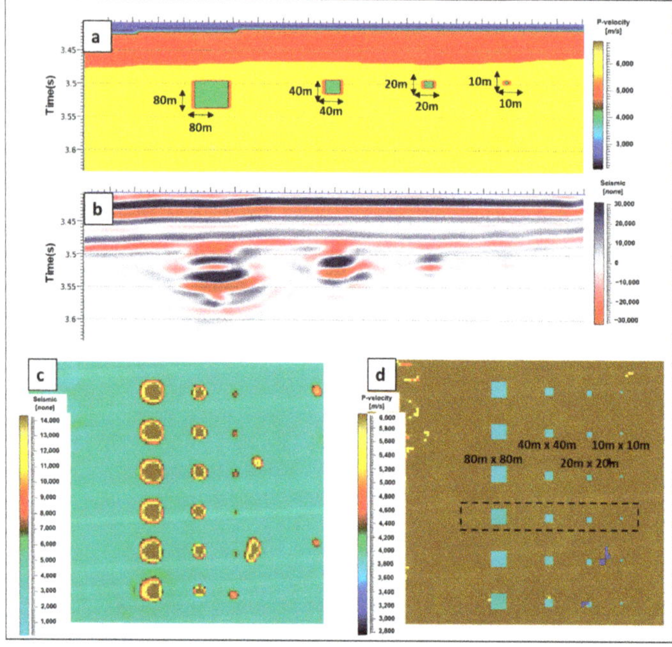

Figure 2. Model of caves of different sizes and depths. Geological model 2: (**a**,**d**) geological models; (**b**) seismic models; (**c**) RMS amplitude map.

Geological model 3: Complicated combinations. This model addresses reservoir lateral connectivity and adjacent caves imaging problems. The analysis of the different reservoir predictions was influenced by the different sizes of the caves and the different distances between them. Figure 3a,d present the view of the depth domain and the map models: Figure 3a shows a section of the model; Figure 3d shows a map of the model. The lateral intervals between the caves are 10 m, 20 m, 80 m, and 120 m. The sizes of caves are 10 m × 10 m, 20 m × 20 m, 80 m × 80 m, 160 m × 160 m, 80 m × 160 m, 20 m × 160 m, and 10 m × 160 m, respectively. All caves have the same velocity of 4000 m/s.

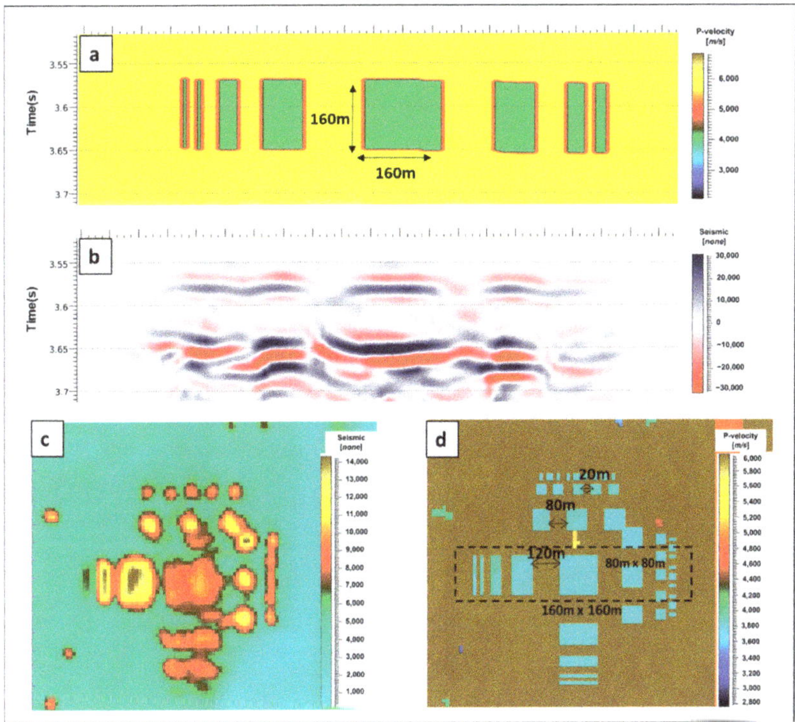

Figure 3. Model of complicated combination caves. Geological model 3: (**a**,**d**) geological models; (**b**) seismic models; (**c**) Seismic RMS amplitude map.

2.2. Method and Workflow

Methods for forward modeling seismic using geological models. The basic steps are as follows: (1) The model is drawn in the display interface based on the principle of non-average quality, layer-shaped technical principles; (2) We accounted for the seismic geometry parameters, the sampling interval, the sampling point, and the initial conditions; (3) The vertical wave dynamic equation can be set as an acoustic wave, an elastic wave, or a viscoelastic wave. The details directly affect the output results, including the decision of whether to use a high-order algorithm, mesh parameters, and boundary bars; (4) Solve the equation. The acoustic wave equation model contains the conversion wave and the S wave effects. It accounts for the distribution of density, the P waves, and the S wave velocity [24,25]. When the wave transitions to the segmentation interface, it produces reflective waves, transparent waves, scattered waves, and wound waves, etc. We used three-dimensional acoustic wave equation modeling [26,27]. In the interest of simulating the actual conditions more precisely, seismic data were used, and the results of each model are shown in Figures 1b, 2b and 3b. In this study, we did not focus on model simulation and seismic reflection instead of the application of pre-stack seismic inversion.

Pre-stack seismic inversion was used to study the applicability of different sets of karst caves in analyzing carbonate caves reservoirs. Synthetic seismic are obtained using 3D full wave equation forward modeling to simulate shot-gather records and then using the Kirchhoff pre-stack depth migration process. As such, our work refers to the main frequency of the actual area in Tahe oil field. A theoretical 30 Hz wavelet was used as the source wavelet, and the max-offset range was 5200 m for forward modeling simulation process [28,29]. We used RockTrace (a CGG pre-stack inversion module) to estimate the elastic parameter (a combination of the P wave velocity, the S wave velocity, and the density) using contained data in multiple-input seismic data (either partial angle stacks or partial offset stacks) to ensure that the entire research process was as close as possible to the actual pre-stack inversion work. As such, we aimed to objectively reflect the effect of applying pre-stack inversion. At each CRP gathering point, the seismic value was recognized as the convolution of a set of reflection coefficients with one or more wavelets. The reflection coefficients were educed from the elastic parameters using Knott–Zoeppritz equations. Converting the angle-dependent seismic amplitudes to elastic parameters provided a quantitative measure of the rock properties.

The RockTrace constrained sparse-peak inversion module created an elastic model based on the superposition of multiple seismic parts (angles or offsets). In each CMP, earthquakes were modeled as a set of reflection coefficients, which convolved with one or more wavelets. The reflection coefficient was derived from the elastic parameter using Knott–Zoeppritz equations (P+P−, P+S−, S+P−, and S+S− reflectivity) or the Aki–Richards approximation. Figure 4 displays the pre-stack constrained sparse-spike inversion which was used in this study. Pre-stack seismic inversion was used in the development of three sets of models. P wave velocity, P wave impedance, S wave impedance, V_p/V_s, and density were produced by inversion. High inversion data quality should be ensured before analysis of the application of pre-stack inversion.

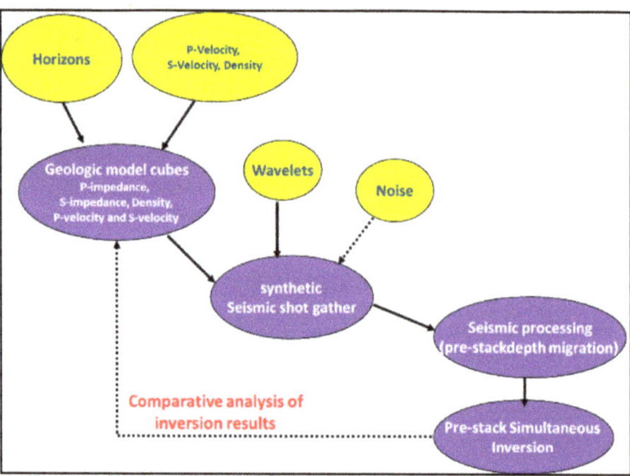

Figure 4. Pre-stack constrained sparse-spike inversion workflow.

3. Results

The carbonate fracture-cave reservoir can be approximately assumed to be an HTI anisotropic medium [28–31], but since the scale of the cave reservoir formed by karst is much larger than fractures, it is approximated as is an isotropic medium. Therefore, the velocity of the geological model and the velocity of the pre-stack inversion can be compared as the best method to evaluate the quality of the inversion data (Figure 5). It is worth noting that the pre-stack seismic inversion can effectively invert the P-wave velocity in the oil and gas production process, and the inverted P-wave velocity is basically consistent with the P-wave velocity of the geological model (Table 2).

P-velocity from geologic model

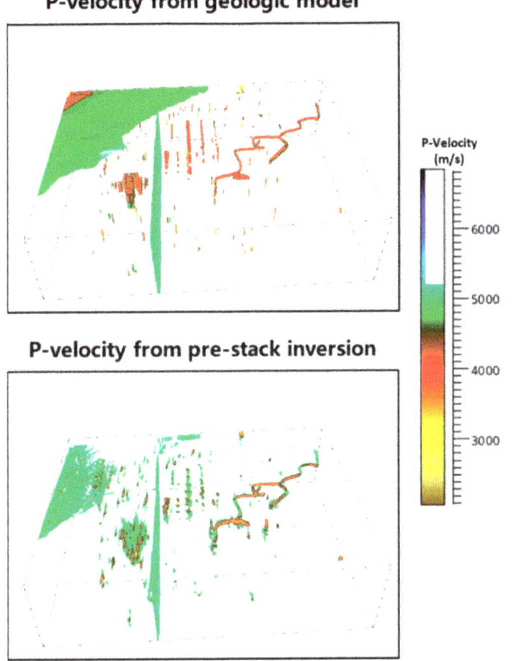

P-velocity from pre-stack inversion

Figure 5. Comparison of P wave velocity of the geological model and that of the pre-stack inversion. The upper picture is the P wave velocity from the geological model; the lower picture is the P wave velocity from the pre-stack inversion.

Table 2. P wave velocity errors between the P wave velocity of the geological model and that of the pre-stack inversion.

	V_p from Geological Model (m/s)	V_p from Pre-Stack Inversion (m/s)	Error
	2800	2800–2900	0–3.57%
	3150	3100–3300	1.58–4.76%
Geological	3550	3400–3700	1.4–4.22%
model 1	4200	4000–4500	4.76 7.14%
	4550	4500–4700	1.09–3.29%
	5150	5100–5300	0.97–2.91%

Figure 6 shows three sections of the pre-stack result for geological model 1: P-impedance, V_p/V_s, and density. P-impedance only shows the position of the caves and the post-stack inversion. V_p/V_s a result only produced by pre-stack inversion, predicts different fillings in cave reservoirs which correspond to the elastic parameters of different fillings. V_P/V_S values can effectively characterize the differences between the different fillings in the caves. The V_p/V_s reflection is strong when the filling velocity is that of gas (2800 m/s), live oil (3150 m/s), oil–water (3550 m/s), or oil (4200 m/s), whereas V_p/V_s has no reflection when the filling is argillaceous (4550 m/s) and calcareous mudstone (5150 m/s). Pre-stack inversion showed significantly improved fluid prediction in comparison with post-stack inversion. The results indicate that pre-stack inversion can obtain basic elastic parameters for fluid prediction. Pre-stack inversion is applicable for fluid prediction in carbonate karst reservoirs.

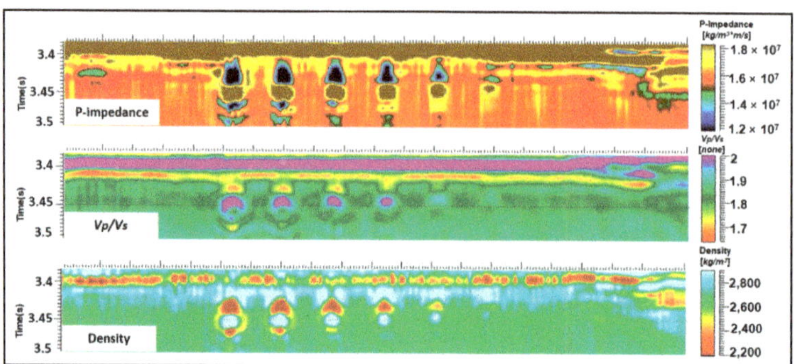

Figure 6. Seismic inverted results of geological model 1: P-impedance prediction the cave reservoir; V_p/V_s ratio effectively predicted in non-oil-bearing caves; density predicted a result between P-impedance and V_p/V_s ratio prediction.

Geological model 2 allowed a comparison between pre-stack inversion and post-stack inversion as there was no fluid filling in the caves.

4. Discussion

4.1. Cave Top Prediction

In Tahe oilfield, the main objective of the current exploration and development is well location design and drilling on top of the cavern reservoir. In Figure 7, the upper picture shows a post-stack inversion section, and the lower picture shows a pre-stack inversion section. Blue indicates the inversion result, and purple indicates the depth of the model's position. The P-impedance of post-stack inversion has been used to predict the position of the top of caves in recent years; however, our results show that pre-stack inversion can more accurately predict the top of the cave. The error is controlled within 1 m, an improvement on the 10 m error of post-stack inversion, which indicates that pre-stack inversion is more effective and applicable for predicting the tops of reservoirs. Pre-stack inversion prediction is a more accurate method for the prediction of the depth of a cave than post-stack inversion prediction. Pre-stack inversion is applicable for the prediction of the full-depth position in carbonate karst reservoirs.

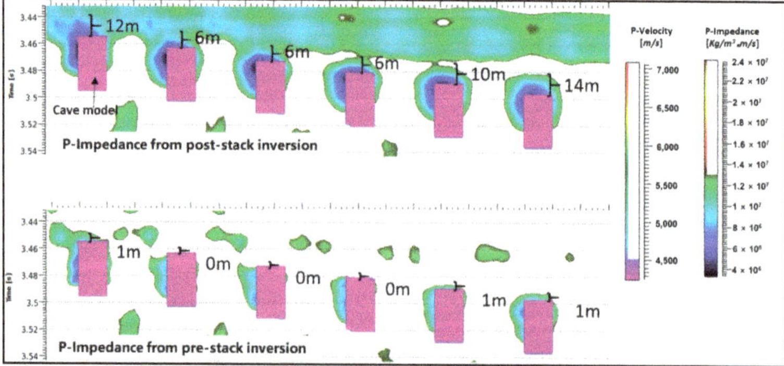

Figure 7. Comparison of P-impedance between post-stack inversion and pre-stack inversion. Purple rectangles indicate models with different distances from the upper surface. Pre-stack inversion shows more effective prediction results.

4.2. Cave Size Prediction

Figure 8 shows four pictures: a geological model map, a seismic attribute map (from acoustic forward modeling), a pre-stack inversion map, and a post-stack inversion map. The four different results show that a 20 m × 20 m cave can be accurately predicted by pre-stack inversion: post-stack inversion predicts a size a little bigger than it is, and the seismic attribute prediction is the least accurate of the four. For caves sized at 40 m × 40 m, the same results were found. Seismic attribute prediction has long been considered as a popular and effective method for characterizing caves in a map. This case shows that pre-stack inversion prediction is the most accurate method for predicting the size of caves compared with seismic data prediction and post-stack inversion prediction. The caves map distribution obtained by post-stack inversion was too different from the geological model: more than 100%. The predictions by pre-stack inversion for the size of the caves are closer to the geological model, especially for smaller caves. Pre-stack inversion is applicable for cave size prediction in carbonate karst reservoirs.

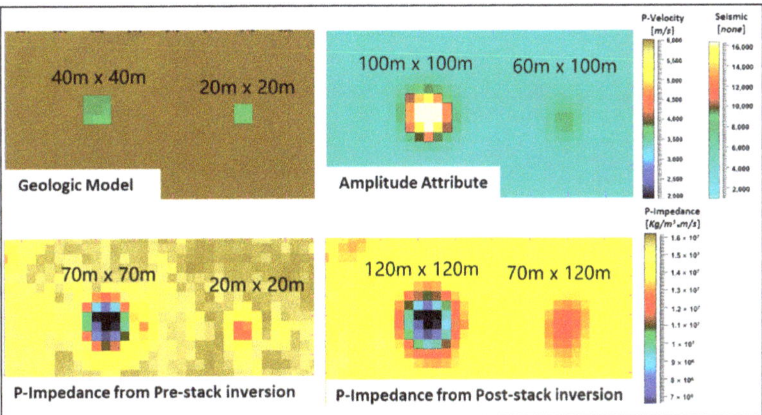

Figure 8. Comparisons of prediction caves size (map): post-stack inversion prediction; pre-stack inversion prediction; seismic attribute prediction. The sizes of the caves predicted by pre-stack inversion are closest to those of the geological model.

Figure 9 shows three pictures for geological model 3: the upper one is the post-stack inversion prediction, the middle one is the pre-stack inversion prediction, and the lower one is the model prediction. The pre-stack inversion result shows layered characteristics for the complicated combination model. The prediction for the top of cave reservoir becomes more accurate with the increase in volume. Pre-stack inversion can more accurately describe the boundary of a single cave in the combination condition than the results of post-stack inversion (shown in Figure 9). Pre-stack inversion displays the shape of these two caves, but the outline is not sufficiently obviously due to the 10 m distance between the two models. The connectivity of the left two caves, measuring 20 m × 20 m and 10 m × 10 m, is poorly defined in post-stack inversion, with little reflection.

Different elastic parameters have different ability to predict reservoirs, the elastic parameters from pre-stack inversion are suitable for the prediction of different types of carbonate reservoirs (shown in Table 3). The density is also relevant in the fluid reservoir, which can help P-impedance and the V_p/V_s in reservoir prediction. P-impedance reflects the position and size of the caves well, and the V_p/V_s and V_p reflects the fluid prediction. Pre-stack inversion is applicable for cave connectivity prediction in carbonate karst reservoirs because it predicted the connectivity of the caves more accurately than post-stack inversion.

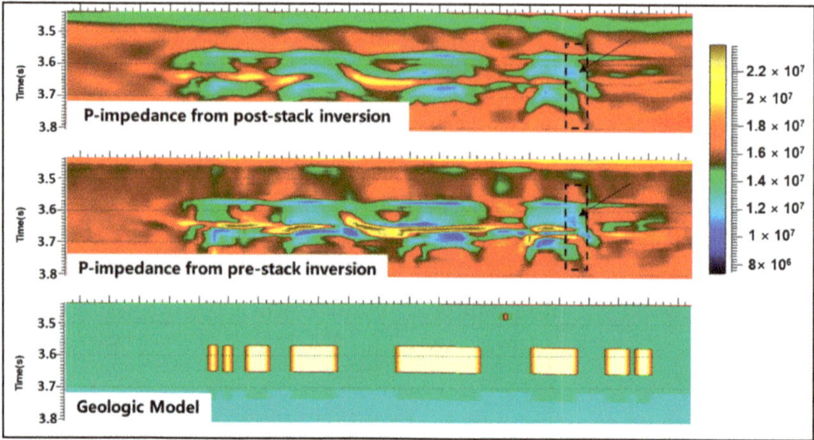

Figure 9. Comparison of P-impedance between post-stack inversion and pre-stack inversion: The black square shows the boundary of one of the caves clearly in the complicated combination; the post-stack inversion result has a lower resolution.

Table 3. Quantitative analysis results of pre-stack inversion.

Model	Model Aim	Analysis
Model 1	Different fillings	Velocity and density related to fluid prediction; velocity 5–10% error with the P wave velocity geological model
Model 2	Top of cave	Predicted accurately
	Different volume	Possible when the caves volume is more than 20 m × 20 m × 20 m; possible to predict the size and position of the cave
Model 3	Complex combination	Layered characterization, predicted the boundary of a single cave

5. Case Study

Pre-stack inversion was studied using seismic data from the Tahe oilfield in the Tarim Basin of Northwestern China. The reservoir is overlaid by a thick layer of clastic rock. The key aim of searching for carbonate reservoirs is to locate karst caves accurately and find water-eroded caves which are filled with oil. As is analyzed in this study, pre-stack inversion was applied for carbonate karst reservoirs.

The application results of pre-stack inversion are shown in Figure 10, where pre-stack inversion P-impedance and V_p/V_s ratio are plotted at the top, and the seismic section is shown at the bottom. The seismic section shows that W1 and W2 are both characterized as cave reservoirs, and the P-impedance result indicates the W1 position and size. W1 is a larger cave than W2, indicating that W1 is a better reservoir than W2; it was hard to judge whether W2 is an oil-bearing reservoir from the P-impedance results. In contrast, the V_p/V_s ratio indicated that W2 is not filled with oil, which is consistent with the known production data.

Figure 11 shows a complicated combination of caves. Seismic data show many strings of beads, and it is hard to distinguish where the caves are. Post-stack inversion indicates about five large caves, as analyzed above, but this indication is difficult for complicated combinations of caves. The pre-stack inversion results show small-scale caves and their boundaries.

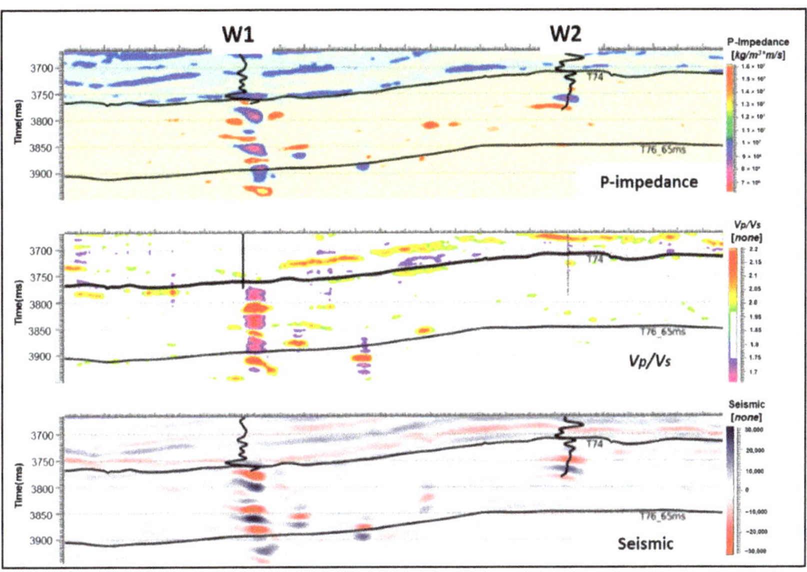

Figure 10. Predicting cave-filling fluids based on pre-stack inversion.

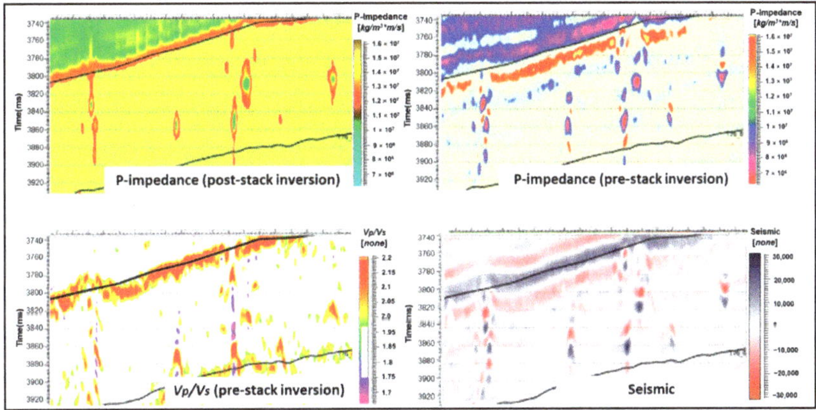

Figure 11. Comparison of P wave impedance profile between post-stack inversion and pre-stack inversion.

6. Conclusions

Through the application of pre-stack inversion in carbonate reservoirs, it can be seen that Pre-stack inversion, as compared to post-stack inversion, provides for more accurate estimates of the depth of carbonate karst reservoirs.

The P wave velocity of the pre-stack inversion is well capable of predicting cave fillings and can perform a quantitative analysis of cave fillings.

Pre-stack inversion has improved cave size prediction in carbonate karst reservoirs to 20 m × 20 m, greatly outperforming seismic data in the prediction of caves.

Pre-stack inversion can be applied for complex reservoirs in carbonate karst reservoirs and is more effective than post-stack inversion at predicting cave connectivity.

Analysis of pre-stack inversion suggests that it is a useful technique for predicting carbonate karst reservoirs in cave reservoir. The comparison results demonstrate that pre-stack inversion improves the prediction of karst reservoir capacities.

Author Contributions: Conceptualization, R.W.; methodology, B.L.; validation, R.W. and B.L.; formal analysis, R.W.; investigation, R.W. and B.L.; resources, R.W.; data curation, B.L.; writing—original draft preparation, B.L.; writing—review and editing, R.W. and B.L.; visualization, B.L.; supervision, R.W.; project administration, R.W.; funding acquisition, R.W. All authors have read and agreed to the published version of the manuscript.

Funding: This research was jointly supported by the National Science and Technology Major Project of China (Beijing) (No. 2017ZX05005004).

Conflicts of Interest: The authors declare no conflict of interest.

References

1. Li, Y.; Yang, J.; Pan, Z.; Meng, S.; Wang, K.; Niu, X. Unconventional Natural Gas Accumulations in Stacked Deposits: A Discussion of Upper Paleozoic Coal-Bearing Strata in the East Margin of the Ordos Basin, China. *Acta Geol. Sin.-Engl. Ed.* **2019**, *93*, 111–129. [CrossRef]
2. Li, Y.; Pan, S.; Ning, S.; Shao, L.; Jing, Z.; Wang, Z. Coal measure metallogeny: Metallogenic system and implication for resource and environment. *Sci. China Earth Sci.* **2022**, *65*, 1211–1228. [CrossRef]
3. Zeng, H.; Wang, G.; Janson, X.; Loucks, R.; Xia, Y.; Xu, L.; Yuan, B. Characterizing seismic bright spots in deeply buried, Ordovician Paleokarst strata, Central Tabei uplift, Tarim Basin, Western China. *Geophysics* **2011**, *76*, B127–B137. [CrossRef]
4. Xu, C.; Di, B.; Wei, J. A physical modeling study of seismic features of karst cave reservoirs in the Tarim Basin, China. *Geophysics* **2016**, *81*, B31–B41. [CrossRef]
5. Jinhua, Y.; Guofa, L.; Yang, L.; Weidong, J. The application of elastic impedance inversion in reservoir prediction at the Jinan area of Tarim Oilfield. *Appl. Geophys.* **2007**, *4*, 201–206. [CrossRef]
6. Feng, X.; Wang, Y.; Wang, X.; Wang, N.; Gao, G.; Zhu, X. The application of high-resolution 3D seismic acquisition techniques for carbonate reservoir characterization in China. *Lead. Edge* **2012**, *31*, 168–179. [CrossRef]
7. Zhang, Y.; Sun, Z.; Jin, Z.; Dong, N.; Chen, Y.; Liu, X. An Amplitude-Based Modeling Method and its Application on the Impedance Inversion in Heterogeneous Paleokarst Carbonate Reservoirs. *Earth Sci. Res.* **2016**, *5*, 199. [CrossRef]
8. Xiang, K.; Han, L.; Hu, Z.; Landa, E. Improving the resolution of impedance inversion in karst systems by incorporating diffraction information: A case study of Tarim Basin, China. *Geophysics* **2020**, *85*, B223–B232. [CrossRef]
9. Zhao, H.; Yin, C.; Li, R.; Xu, F.; Chen, G.; Li, X.; Fu, L. Design method of gun offset based on maximizing the energy efficiency of the receiver receiving illumination. *Oil Geophys. Geophys.* **2011**, *46*, 333–338.
10. Biot, M.A. Generalized Theory of Acoustic Propagation in Porous Dissipative Media. *J. Acoust. Soc. Am.* **1962**, *34*, 1254–1264. [CrossRef]
11. Zhang, Y.; Sun, S.Z.; Yang, H.; Wang, H.; Han, J.; Gao, H.; Luo, C.; Jing, B. Pre-stack inversion for caved carbonate reservoir prediction: A case study from Tarim Basin, China. *Pet. Sci.* **2011**, *8*, 415–421. [CrossRef]
12. Sun, X.; Sam, Z.S.; Xie, H. Nonstationary sparsity-constrained seismic deconvolution. *Appl. Geophys.* **2014**, *11*, 459–467. [CrossRef]
13. Zhan, G.; Pestana, R.C.; Stoffa, P.L. Decoupled equations for reverse time migration in tilted transversely isotropic media. *Geophysics* **2012**, *77*, 37–45. [CrossRef]
14. Zhang, Y.Y.; Sun, Z.D.; Han, J.F.; Wang, H.Y.; Fan, C.Y. Fluid mapping in deeply buried Ordovician paleokarst reservoirs in the Tarim Basin, western China. *Geofluids* **2016**, *16*, 421–433. [CrossRef]
15. Sun, S.Z.; Yang, P.; Liu, L.; Sun, X.; Liu, Z.; Zhang, Y. Ultimate use of prestack seismic data: Integration of rock physics, amplitude-preserved processing, and elastic inversion. *Lead. Edge* **2015**, *34*, 308–314. [CrossRef]
16. Wang, X.W.; Lei, L.; Liu, W. Seismic data processing techniques of carbonate rocks in Tarim Basin. *Lithol. Reserv.* **2008**, *20*, 109–112.
17. Sun, S.Z.; Yang, H.; Zhang, Y.; Han, J.; Wang, D.; Sun, W.; Jiang, S. The application of amplitude-preserved processing and migration for carbonate reservoir prediction in the Tarim Basin, China. *Pet. Sci.* **2011**, *8*, 406–414. [CrossRef]
18. Zhang, Y.; Sun, Z.; Jin, Z.; Fan, C. The Comparison between Full-Stack Data and Pure P-Wave Data on Deeply Buried Ordovician Paleokarst Reservoir Prediction. *Earth Sci. Res.* **2016**, *5*, 57–59. [CrossRef]
19. Xu, C.; Di, B.; Wei, J. Influence of Reservoir Size on AVO Characteristics of Reservoirs at Seismic Scale. In Proceedings of the 76th EAGE Conference and Exhibition 2014, Amsterdam, The Netherlands, 16–19 June 2014. [CrossRef]
20. Wang, B.L.; Yin, X.Y.; Zhang, F.C. Elastic impedance inversion and its application. *Prog. Geophys.* **2005**, *20*, 89–92.
21. Zhang, Y.; Sun, Z.; Fan, C. An Iterative AVO Inversion Workflow for S-wave Improvement. *Geology* **2013**, *10*, 1–3.
22. Gan, L.D.; Zhao, B.L.; Du, W.H.; Li, L.G. The potential analysis of elastic impedance in the lithology and fluid prediction. *Geophys. Prospect. Pet.* **2005**, *44*, 504–508.
23. Wang, H.; Sun, S.Z.; Yang, H. Velocity prediction models evaluation and permeability prediction for fractured and caved carbonate reservoir: From theory to case study. In *Society of Exploration Geophysicists SEG Technical Program Expanded Abstracts 2009*; Society of Exploration Geophysicists: Houston, TX, USA, 2009; pp. 2194–2198.

24. Etgen, J.T.; Brandsberg-Dahl, S. The pseudo-analytical method: Application of pseudo-Laplacians to acoustic and acoustic anisotropic wave propagation. In *Society of Exploration Geophysicists SEG Technical Program Expanded Abstracts 2009*; Society of Exploration Geophysicists: Houston, TX, USA, 2009; pp. 2252–2256.

25. Huang, J.Q.; Li, Z.C. Modeling and reverse time migration of pure quasi-P-waves in complex TI media with a low-rank decomposition. *Chin. J. Geophys.* **2017**, *60*, 704–721.

26. Xu, S.; Zhou, H. Accurate simulations of pure quasi-P-waves in complex anisotropic media. *Geophys. J. Soc. Explor. Geophys.* **2014**, *79*, 341–348. [CrossRef]

27. Wang, Y.; Mu, P.; Cai, W.; Wang, P.; Gui, Z. Optimization of five-diagonal compact difference scheme and numerical simulation of two-dimensional acoustic wave propagation wave equation. *Geophys. Prospect. Pet.* **2019**, *58*, 487–498.

28. Zhang, J.-M.; He, B.-S.; Tang, H.-G. Pure quasi-p wave equation and numerical solution in 3d tti media. *Appl. Geophys.* **2017**, *14*, 125–132. [CrossRef]

29. Li, Q.; Wu, G.; Duan, P. Quasi-regular grid high-order finite difference method for heterogeneous elastic wave field simulation. *Oil Geophys. Prospect.* **2019**, *54*, 540–550.

30. Chu, C.; Macy, B.K.; Anno, P.D. Approximation of pure acoustic seismic wave propagation in TTI media. *Geophysics* **2011**, *76*, WB97–WB107. [CrossRef]

31. Du, Q.Z.; Guo, C.F.; Gong, X.F. Hybrid PS/FD numerical simulation and stability analysis of pure P-wave propagation in VTI media. *Chin. J. Geophys.* **2015**, *58*, 287–301. [CrossRef]

Article

Inversion Study on Parameters of Cascade Coexisting Gas-Bearing Reservoirs in Huainan Coal Measures

Baiping Chen [1], Bo Liu [2], Yunfei Du [1], Guoqi Dong [1], Chen Wang [1], Zichang Wang [1], Ran Wang [1] and Fan Cui [1,*]

1 College of Geosciences and Surveying Engineering, China University of Mining and Technology-Beijing, Ding No. 11 Xueyuan Road, Haidian District, Beijing 100083, China
2 School of Earth Science and Engineering, Hebei University of Engineering, No. 19 Tai Chi Road, Economic and Technological Development Zone, Handan 056083, China
* Correspondence: cuifan_cumtb@126.com

Abstract: The prediction and development of three gases, mainly coalbed methane, shale gas, and tight sandstone gas, in the Huainan coal measures of China, has been the focus of local coal mines. However, due to the overlapping and coexisting characteristics of the three gas reservoirs in Huainan coal measure strata, it is challenging to develop the three gas. The coal mine has been creating a single pool for a long time, resulting in the severe waste of other gas resources in developing the gas-bearing resources in the coal measure strata. The gas-containing reservoir is predicted based on geological, seismic, and logging in Huainan Mining. In addition, determining the excellent area for reference for the development of three gas resources. First, using logging data, mathematical–statistical methods are used to analyze the physical parameters of gas-bearing reservoirs in multi-layered stacked coal seams. Then, based on the theory of prestack seismic inversion, parameters, such as the impedance of P-wave, the ratio of P-wave velocity and S-wave, Lamé constant, Young's modulus, and Poisson's ratio and lithological distribution, are obtained for the whole area. The gas-bearing information of the reservoir is received by the statistics and equation of the parameter intersection diagram and is closely related to exploration and development. Finally, the paper synthetically predicts the most favorable area of the gas-bearing reservoir in the study area. The prediction results are compared with the actual results of coalbed methane content in the existing extraction wells, proving that the method is feasible and can provide the basis for the deployment and development of the well location.

Keywords: coalbed methane (CBM); prestack seismic inversion; brittle index; gas-bearing property

Citation: Chen, B.; Liu, B.; Du, Y.; Dong, G.; Wang, C.; Wang, Z.; Wang, R.; Cui, F. Inversion Study on Parameters of Cascade Coexisting Gas-Bearing Reservoirs in Huainan Coal Measures. *Energies* **2022**, *15*, 6208. https://doi.org/10.3390/en15176208

Academic Editor: Reza Rezaee

Received: 30 June 2022
Accepted: 18 August 2022
Published: 26 August 2022

Publisher's Note: MDPI stays neutral with regard to jurisdictional claims in published maps and institutional affiliations.

1. Introduction

Currently, the three gases of coal measures in unconventional natural gas, mainly coalbed methane, tight sandstone gas, and shale gas, have been paid attention to and developed in many countries [1]. China is rich in coal resources, ranking third in the world, and has large gas reservoirs in coal measure strata. At the same time, the demand for natural gas is huge, and the natural gas gap is expected to reach 200 billion cubic meters in 2030 [2]. Many researchers have also done a lot of work on predicting and developing coalbed methane in the Huainan area of China [3,4]. However, the gas-bearing reservoirs of coal measure strata in the Huainan area are characterized by many types and superpositions, leading to the waste of other gas resources in developing coal bed methane [5]. However, the characteristics of superimposed and coexisting gas-bearing reservoirs in Huainan coal measures bring difficulties to build and significant challenges to reservoir prediction in the early stage.

The three gas reservoirs (coalbed methane (CBM), shale gas, tight sandstone gas) in the Huainan coal measures are basically in the Carboniferous–Permian coal measures, with geological characteristics, such as diverse natural gas occurrence, lithology variation, and overlapping of reservoir and caprock, the coexistence of reservoir fluid pressure system, and significant difference in mechanical properties [6]. Therefore, in the prediction of three different gas reservoirs, not only the lithology, buried depth, thickness, gas-bearing characteristics, and other geological factors should be considered, but also the overall sequence stratigraphic framework characteristics of coal measure strata and the energy and pressure differences among superimposed gas bearing systems should be considered. Therefore, it is necessary to integrate many kinds of information, such as reservoir lithology, reservoir buried depth, reservoir thickness, and gas-bearing characteristics, to predict the type of gas reservoir accurately. Currently, 3D seismic prestack inversion is a common method for oil and gas reservoir prediction, which many researchers have studied. For example, Zhou (2017) used the accurate Zoeppritz equation to realize the inversion of two elastic parameters, Young's modulus and Poisson's ratio, to effectively predict the location of shale gas reservoir [7]. Zhang (2020) proposed they would obtain reservoir parameters of porosity, sand index, and density directly from prestack seismic data through a petrophysical model and prestack amplitude variation with offset (AVO) inversion [8]. The 3D seismic prestack inversion first uses logging data for petrophysical analysis to find the rules and differences of elastic physical parameters in gas-bearing and non-gas-bearing reservoirs. It then establishes the interpretation version, which is reasonably applied to prestack seismic inversion to obtain the reservoir lithology and elastic parameters in the whole area. It then predicts the lithology and fluid information of the reservoir. The intersection result of $\lambda \cdot \rho$ and $\mu \cdot \rho$ can be used to indicate the gas-bearing property of the reservoir. At the same time, the brittleness index represents the brittleness of the reservoir, which is conducive to the generation and fracturing of reservoir fractures [9,10]. For example, Yasin et al. (2021) predict the shale reservoir in the Longgamasi area of China in the Sichuan Basin, China with a brittle template and have achieved successful results [11].

In summary, reservoir parameter inversion has been widely used in reservoir prediction. Currently, the research on reservoir parameters of coal measure formation gas mainly focuses on CBM or single coal seam, and there are few studies on superimposed and coexisting three gas. Given the superposition and co-existence characteristics of the complex three gas reservoirs in Huainan coal measure strata of China, this paper comprehensively studies the systematic reservoir parameters of the three gases. It predicts the favorable gas-bearing areas, which play a positive role in the joint exploitation of the three gases and the well location deployment in the Huainan coal mine.

2. Geologic Setting

The study area is located at Zhangji Coal Mine, 20 km west of Fengtai County, Huainan City, Anhui Province, China (Figure 1). The study area is a fully concealed coal-bearing area, and the buried depth is below 600 m, which is lower than the weathering zone of CBM. The coal measure strata, Carboniferous–Permian in age, are covered with a considerable thickness of loose sediments of younger strata, of Tertiary–Quaternary in age, and overlie the Ordovician strata. In more detail, the coal-bearing strata are Taiyuan of Upper Carboniferous, Shanxi Formation of Permian, Lower Shihezi Formation of Permian, and Upper Shihezi Formation of Permian. The target beds of this study are coal-bearing strata of the Permian Shanxi Formation and Upper and Lower Shihezi Formation, mainly delta deposits. The main coal seams that can be mined are 13-1$_{\#}$, 11-2$_{\#}$, 8$_{\#}$, 6$_{\#}$, and 1$_{\#}$ from top to bottom, with an average thickness of 5.7 m, 2.4 m, 3 m, 0.8 m, and 7.2 m, respectively. The coal seam roof is mainly composed of mudstone and sandstone, followed by siltstone and sandstone. The base is mainly composed of mudstone and sandstone and partially consists of fine silt and fine sandstone. Faults developed in the south and north edges of the study area, and the number of fractures in the inner minefield is significant but with a small offset. The storage environments and genesis of the three unconventional natural gas in the study area are shown in Table 1.

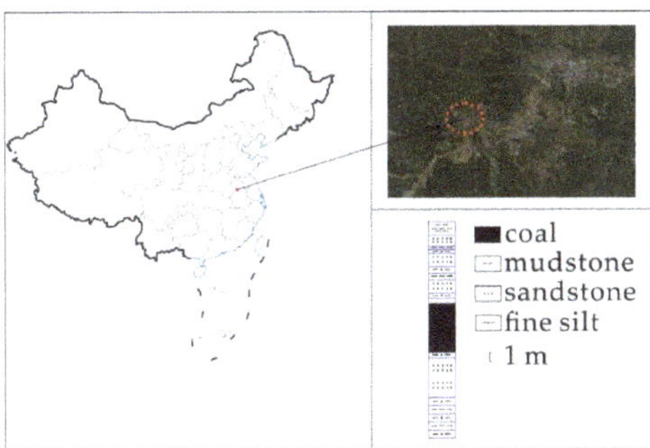

Figure 1. Study area location and lithology histogram.

Table 1. The storage environment and genesis of the three kinds of unconventional natural gas.

	Shale Gas	CBM	Tight Sandstone Gas
Reservoir forming conditions	In situ-generation, in situ-storage, and in situ-preservation	In situ-generation, in situ-storage, and in situ-preservation	Reasonable combination of source, reservoir, and caprock
Definition	It mainly focuses on the natural gas in mud/shale series in adsorption and free state	The natural gas mainly accumulates in coal measure strata in the adsorption state	Under the influence of buoyancy, it focuses on the natural gas at the top of the reservoir
Genetic type	Origin of the thermal evolution of organic matter	Organic matter is formed by thermal evolution and biogenesis	Thermal expansion of organic matter and cracking of crude oil
Occurrence state	20–85% is adsorption; the rest is free and water-soluble	More than 85% of them are adsorbed, and the rest are free and water soluble	The top high points of various traps do not consider the influence factors of adsorption
Reservoir conditions	Characteristics of low porosity and permeability	Dual porosity (matrix and cleat system) Φ: 1–5%; K: 0.5–5.0 md	(1) Low permeability: Φ: 8–20%; K: 0.1–50 md

This study is aimed at gas-bearing reservoirs in coal measure strata. Previous studies believe that AVO has a good prediction effect on gas-bearing reservoirs, so CRP channel gathering seismic data are used to carry out the analysis [12,13]. The frequency band range of this channel is 20–78 Hz, the central frequency is 40 Hz, the offset range is about 0–600 m, and the maximum incidence angle of the target layer is 35 degrees (Figure 2).

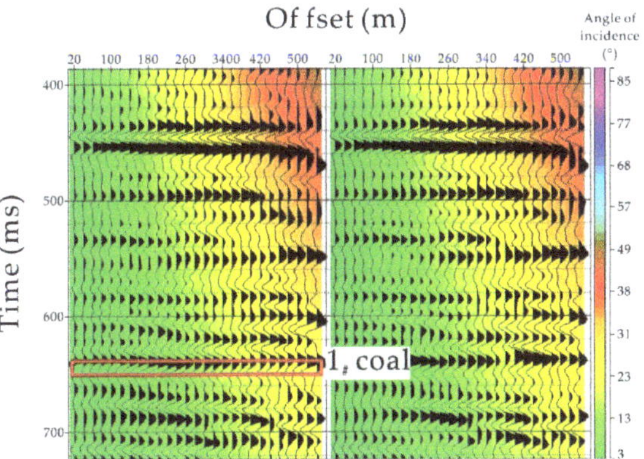

Figure 2. The 3D CRP seismic data profile in the study area.

3. Data and Methods

3.1. Petrophysical Analysis

Coal measure strata are formed in a genetically linked marine–terrestrial or continental sedimentary environment. With this background, CBM, tight sandstone gas, and shale gas reservoirs are layered and co-existed. Their reservoir–cap relationship is complex, and there are conversion changes. There are apparent differences in the types of layer rocks, so a unified petrophysical analysis of the three reservoirs is required [14].

The petrophysical analysis is mainly based on the data of parameter well XX − 1 drilled in 2017 in the study area. This well contains 13-1$_#$, 11-2$_#$, 8$_#$, 6$_#$, and 1$_#$ coal layers in the Permian Shanxi Formation. The well was analyzed for reservoir physical parameters by electrical, radioactive, acoustic, and density logging curves, and lithological rules were used to name the formation based on reservoir physical parameters, classifying the lithology into coal rock, tight sandstone gas reservoirs, dry sandstone, shale gas reservoirs, and mudstone. All of the coal rocks contain gas, and the coal rocks can be approximated as coalbed methane reservoirs (Figure 3).

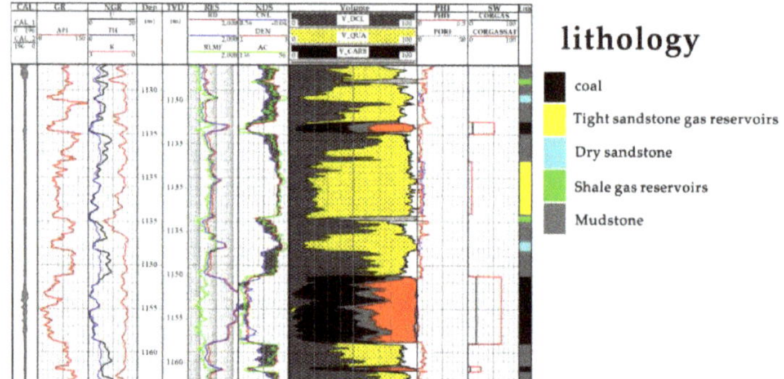

Figure 3. XX − 1 logging interpretation results (Column CAL: borehole diameter curve; Column GR: Gamma curve; Column NGR: energy spectrum curve; Column Dep: depth; Column RES: resistivity curve; Column NDS: Compensated neutron curve, density curve, and acoustic curve; Column Volume: volume of rock physics; Column PHI: porosity curve; Column SW: saturation curve; Column Lith: Lithology).

3.1.1. Relationship between Reservoir Elastic Parameters and Lithology

Reservoir elastic parameters mainly refer to parameters, such as P-wave impedance (Pimp), P-wave velocity (V_p), S-wave velocity (V_s), density, and P-wave to S-wave velocity ratio (V_p/V_s) related to 3D seismic data. These elastic parameters can be directly obtained from the logged data. The ability of adjustable parameters to distinguish lithology is the key to reservoir lithology inversion [15]. In the Carboniferous–Permian coal measure strata, there are many gas-bearing reservoirs, so the intersection of elastic parameters is used to show the distribution law between elasticity and lithology (Figure 4).

3.1.2. Relationship between Reservoir Elastic Parameters and Gas-Bearing Properties

CBM, tight sandstone gas, and shale gas reservoirs have severe overlaps in elastic parameters, and it is challenging to predict gas-bearing reservoirs using conventional flexible parameters. In this study, through the $\lambda \cdot \rho - \mu \cdot \rho$ intersection test and analysis, the recognition effect of gas-bearing reservoirs in the study area is obvious. The distribution is relatively concentrated, which can be used to identify all gas-bearing reservoirs in coal measure strata (Figure 5).

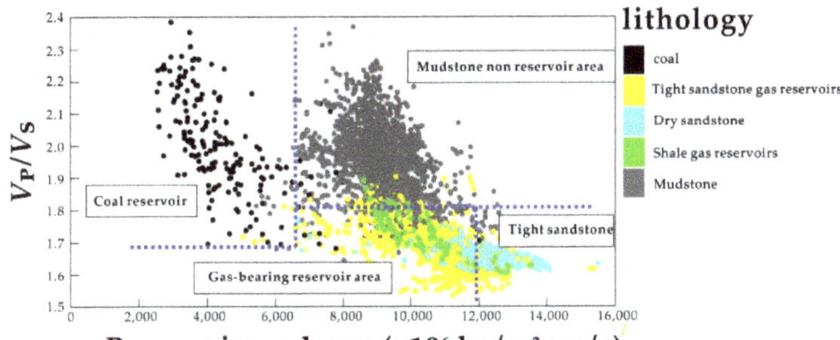

Figure 4. The cross plot of P-wave impedance-P-wave and S-wave velocity ratio after the correction of petrophysics (coal bed is relatively low wave impedance, high P-wave, and S-wave velocity ratio; sandstone gas bed is medium-high wave impedance, low P-wave, and S-wave velocity ratio; dry sandstone bed is relatively high wave impedance, low P-wave, and S-wave velocity ratio; shale gas reservoir is medium-low P-wave impedance, medium-low P-wave velocity ratio).

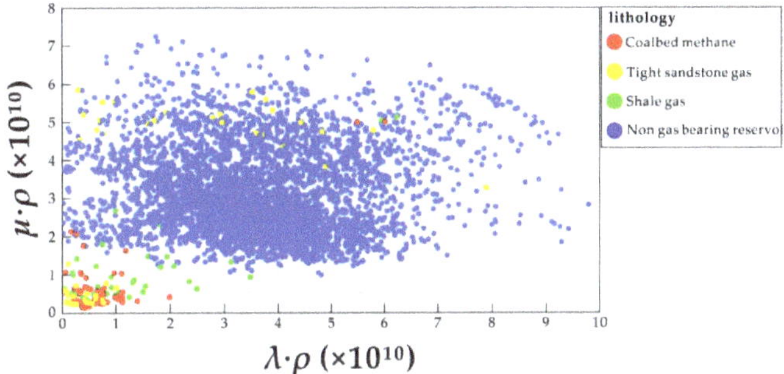

Figure 5. Cross plot of $\lambda\cdot\rho$-$\mu\cdot\rho$.

3.1.3. The Relationship between Elasticity and Rock Brittleness

The definition of rock brittleness includes two critical parameters: Poisson's ratio and Young's modulus. In the description of the rock mechanics method, the brittleness coefficient is defined by Equations (1)–(3) [16–18]:

$$BI = (YM_{BRIT} + PR_{BRIT})/2, \tag{1}$$

$$YM_{BRIT} = (YMS - YMS_{min})/(YMS_{max} - YM_{Smin}), \tag{2}$$

$$PR_{BRIT} = (PR - PR_{max})/(PR_{min} - PR_{max}), \tag{3}$$

In the above Equation: YMS means measured Young's modulus, MPa; PR means measured Poisson's ratio, dimensionless; YM_{BRIT} means normalized Young's modulus, dimensionless; PR_{BRIT} means uniformized Poisson's ratio, dimensionless; BI stands for brittleness index.

XX − 1 Well Poisson's ratio and Young's modulus (static) can be calculated by V_p, V_s, density, and finally, use Equations (1)–(3) to obtain the brittleness index, and give the largest, minimum brittle boundary point feature parameters (Figures 6 and 7). The trend of the most significant friable boundary is high Young's modulus, low Poisson's ratio, and the characteristics of high brittleness and low plasticity. This type of reservoir is conducive to fracturing development; the trend of the minor brittle boundary is low Young's modulus, high Poisson's ratio, and has the characteristics of low brittleness and high plasticity, which is suitable as the cap layer of the reservoir.

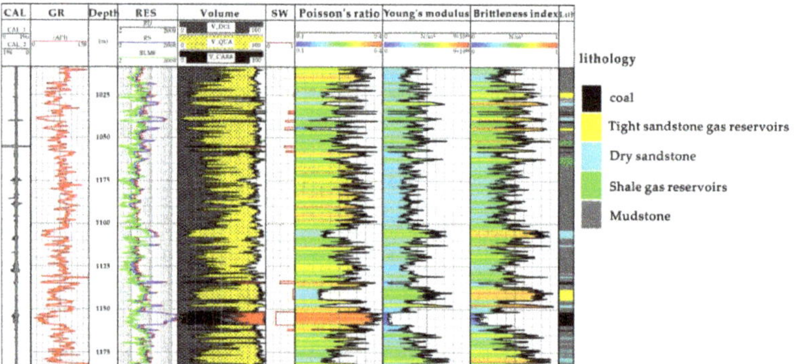

Figure 6. XX − 1 well elasticity and rock brittleness diagram (coal bed gas reservoir shows high Poisson's ratio, low Young's modulus, and low brittleness index; tight sandstone gas reservoir shows low Poisson's ratio, high Young's modulus, and high brittleness index; shale gas reservoir shows middle Poisson's ratio, middle high Young's modulus, and middle high brittleness index).

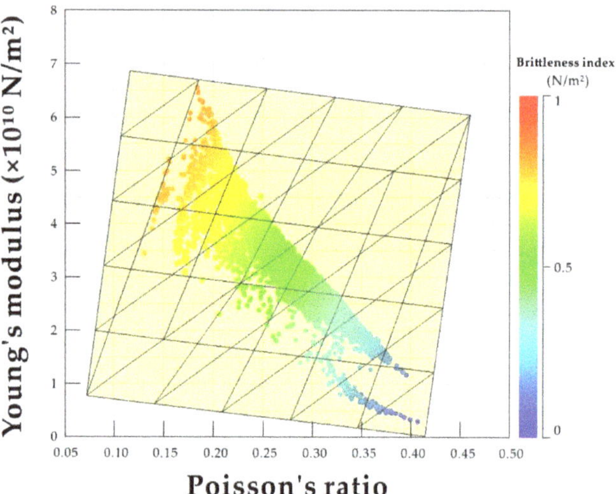

Figure 7. Cross plot of Poisson's ratio and Young's modulus.

3.2. Simultaneous Prestack Inversion

3.2.1. AVO Forward Analysis

AVO response characteristics are affected by the reservoir's longitudinal and transverse wave velocity and density. The CBM content affects the coal reservoir's longitudinal and transverse wave velocity and density. Many researchers have done a lot of research on the forward modeling of conventional gas reservoirs and clearly understand the characteristics of AVO [19–21]. The petrophysical analysis in the study area shows that the elastic parameters of shale gas reservoirs are layered with tight sandstone gas reservoirs, so the AVO types of the two are similar. Considering that the elasticity and physical properties of coal quality are quite different in different regions, this paper carries out AVO forward modeling of the CBM reservoirs in the study area to evaluate the AVO characteristics of the CBM reservoirs. This paper designs 6 different gas saturation models for CBM reservoirs: 0%, 20%, 40%, 60%, 80%, 100% to verify the relationship between CBM content and AVO characteristics. The results show that the intercept attribute has a more obvious response to the enrichment degree of CBM, indicating that the post-stack seismic data will be affected by the large incident angle seismic data when used in reservoir inversion. Therefore, it is necessary to use prestack seismic data to conduct gas-bearing reservoir prediction research (Figure 8).

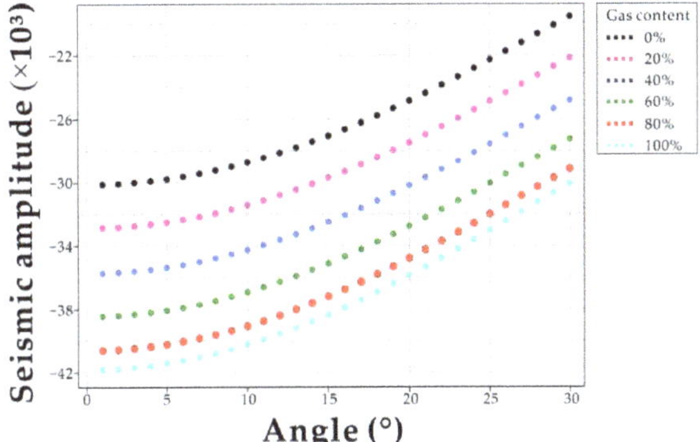

Figure 8. The amplitude of coal roof with different gas content in the forward model varies with the offset (intercept is all negative and decreases with the increase in gas content (absolute value increases), and the degree of decrease becomes smaller. The amplitude increases with the increase in incident angle (offset) (the total value decreases). Gradients are positive and do not change significantly with gas content.

3.2.2. Prestack Gathers Optimization Processing

Reservoir petrophysical research and forward AVO analysis show that the research of gas-bearing reservoirs need to obtain prestack inversion results, such as shear wave impedance, Poisson's ratio, and Lame constant. The authenticity and accuracy of seismic data directly affect the results of elastic inversion and the accuracy of reservoir prediction. Therefore, this study needs higher requirements for amplitude preservation and fidelity of CRP data. The processing of prestack CRP gathers requires strict amplitude preservation and AVO preservation features. In this study, only noise suppression is performed on seismic gather data. It can be seen that this method reduces noise interference, effectively improves the signal-to-noise ratio, and ensures the quality of the subsequent prestack simultaneous inversion of seismic data (Figure 9).

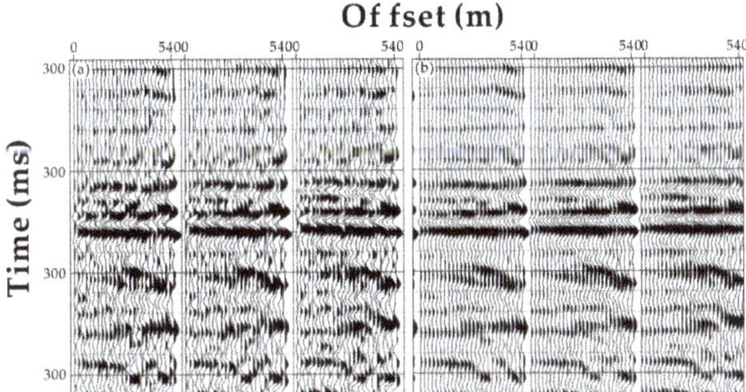

Figure 9. Comparison of 3D seismic tracking data before and after optimization process. (**a**) is the seismic profile before optimization. (**b**) is the optimized seismic profile.

Simultaneous prestack inversion requires multiple partially superimposed data volumes with a high signal-to-noise ratio as input. To ensure the stability of the inversion result and a high signal-to-noise ratio, generally, at least 3 to 5 angle gathering superimposed profiles are required as input data. Due to the low data coverage of the measured channels in this study, the collected data are divided into three-angle parts, superimposed according to the maximum incident angle. Namely, the angle part superimposes 1~12 degrees, 12~24 degrees, and 24~35 degrees (Figure 10).

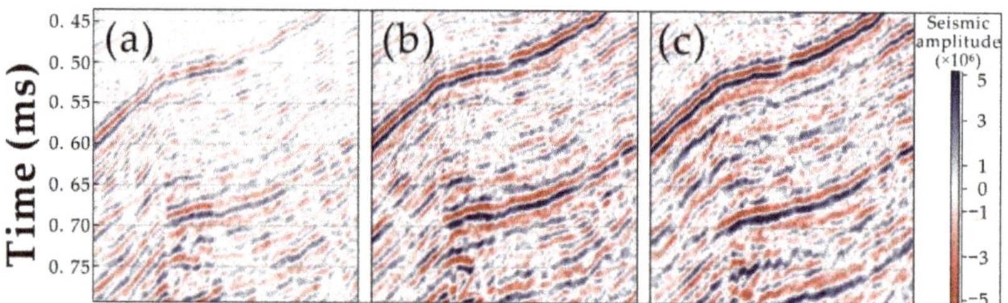

Figure 10. Partial overlay of angle data ((**a**) Partial overlay of angle 1–12°, (**b**) Partial overlay of angle 12–24°, (**c**) Partial overlay of angle 24–35°).

3.2.3. Prestack Inversion

Prestack inversion is also called simultaneous inversion. P-wave velocity and S-wave velocity are calculated together with density [22]. This inversion is performed on the prestack seismic data (incident angle superimposed data or offset superimposed data). Finally, the reservoir parameters, such as compressional wave impedance, density, $\lambda \cdot \rho$, $\mu \cdot \rho$, etc., are obtained. The fluid category is of great significance, and the method is mainly based on the elastic impedance equation (Equations (4)) [23,24].

$$R_{PP}(\theta) = \frac{\Delta\lambda}{\lambda}\left[\frac{1}{4} - \frac{1}{2}\left(\frac{v_s}{v_p}\right)^2\right]\left(sec^2\theta\right) + \frac{\Delta\mu}{\mu}\left(\frac{v_s}{v_p}\right)^2\left(\frac{1}{2}sec^2\theta - 2sin^2\theta\right) + [\frac{1}{2} - \frac{1}{4}sec^2\theta]\frac{\Delta\rho}{\rho} \qquad (4)$$

Among them: V_p is the longitudinal wave velocity, V_s is the transverse wave velocity, and ρ is the density.

The prestack inversion part mainly includes the data needed in angle-stacked seismic data, wavelet, and low-frequency trend model. In the process of inversion, they used a constrained light pulse global optimization method for different incidence angles of multiple seismic data volumes simultaneously inversion. This method objectively uses pre-seismic data information to ensure the reliability of the anti-elastic parameters. The inversion process is shown in Figure 11.

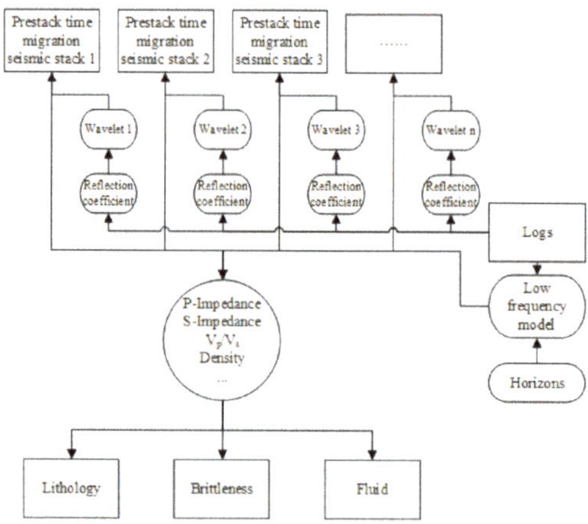

Figure 11. Prestack simultaneous inversion flow.

Obtain P-wave impedance body, P-wave velocity ratio body, density body, $\lambda \cdot \rho$, $\mu \cdot \rho$, Poisson's ratio, and other data bodies (Figure 12). The emphasis of these elastic parameters in geological interpretation is different. The petrophysical analysis shows that the P-wave impedance and P-wave velocity ratios are related to lithology, $\lambda \cdot \rho$ and $\mu \cdot \rho$ are related to gas-bearing properties, and Poisson wave and Young's modulus (static) are related to brittleness index. This is the subsequent geological explanation that provides a basis.

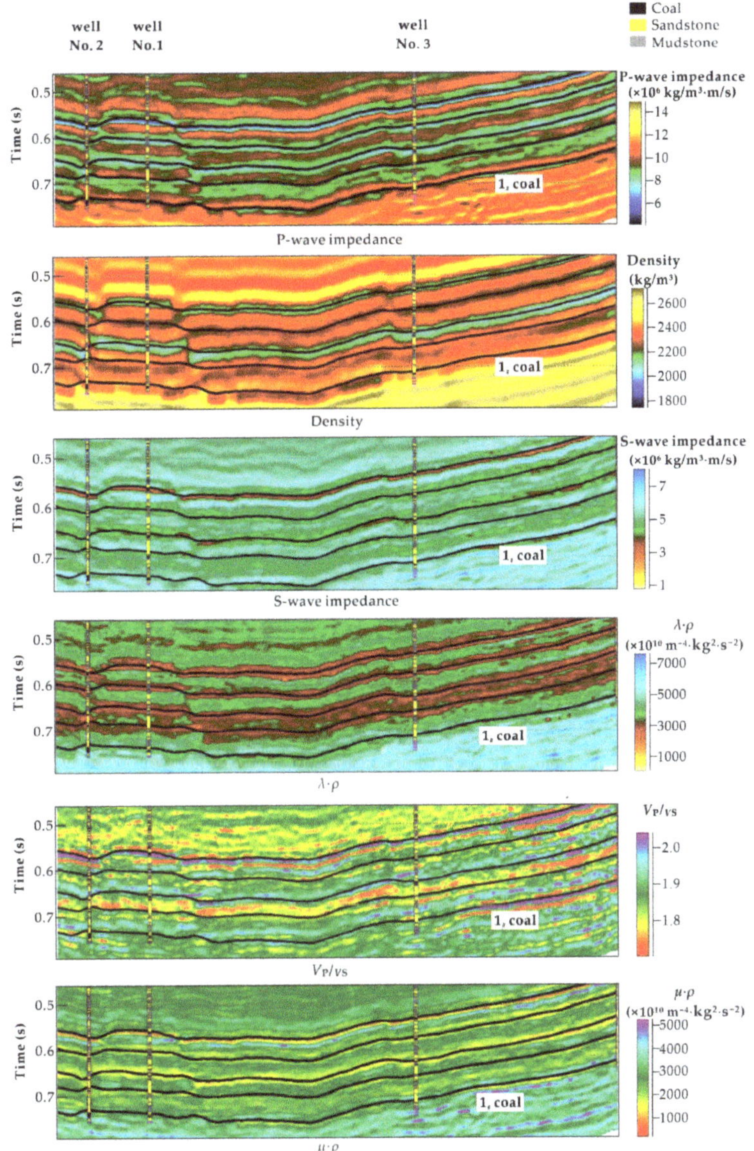

Figure 12. Parameter profiles of P-wave impedance, P-wave velocity ratio, density, $\lambda \cdot \rho$, $\mu \cdot \rho$, and Poisson's ratio.

3.3. Analysis of Brittleness Index

Log brittleness index results show that Poisson's ratio and Young's modulus jointly determine the brittleness index. Based on Equations (1)–(3), the brittleness index is calculated using the inverted Poisson's ratio and Young's modulus (Figure 13). A high value indicates that the formation is brittle and vice versa.

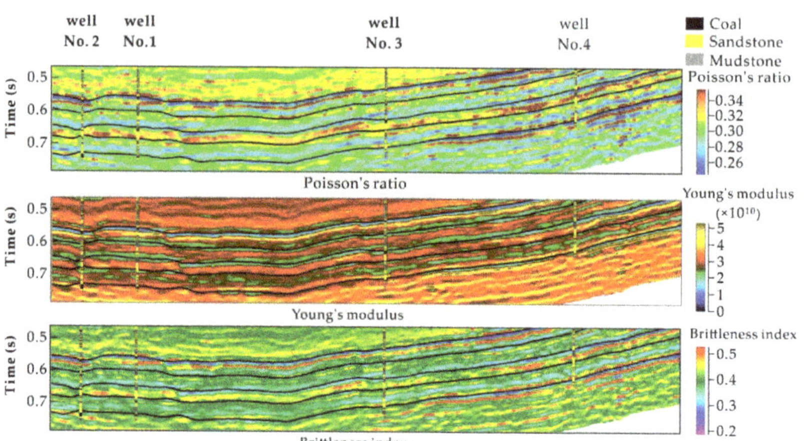

Figure 13. Profiles Poisson's ratio, Young's modulus, Brittleness index inversion.

3.4. Probability Analysis of Lithologic Fluid

The probabilistic analysis process of lithologic fluid is as follows: firstly, the reservoir parameter bodies, such as P-wave impedance, P-wave velocity ratio, density, $\lambda \cdot \rho$, $\mu \cdot \rho$, Poisson's ratio, etc., are obtained by simultaneous prestack inversion, and then based on the results of rock physics analysis, probabilistic analysis of the conversion results in the lithology and fluid spatial distribution [25]. The traditional interpretation process uses the method of elastic parameter cut-off value. This method cannot accurately describe the uncertainty and has large errors. However, the probabilistic volume analysis technology of lithological fluid overcomes these problems. It combines the deterministic petrophysical relationship with statistics, and statistical techniques are used to describe the uncertainty and spatial changes transmitted during the transformation of rock properties [26]. Through lithology statistics and random simulation of the well point data, the response range and quantitative probability distribution function of different lithologies or fluids corresponding to the inversion data volume are established, and the posterior probability of each lithofacies (fluid) is calculated. Finally, the inversion parameters of the reservoir are converted into lithologic bodies and gas-bearing reservoir probabilistic bodies under seismic resolution by two-dimensional table transformation [27]. This paper uses $\lambda \cdot \rho$, $\mu \cdot \rho$ parameters to predict the gas-bearing properties of coal-measure formations and predicts the lithology based on the parameters of P-wave impedance and P-wave velocity ratio. The superimposed study of lithology and gas-bearing properties will be beneficial for storage. The warm-red color in the figure represents the high gas-bearing probability developed in different lithologies. Among them, coal bed methane reservoirs are the main gas-bearing reservoirs of coal-measure strata, followed by tight sandstone gas reservoirs and shale gas reservoirs. There are a few layers that are consistent with the actual logging interpretation (Figure 14).

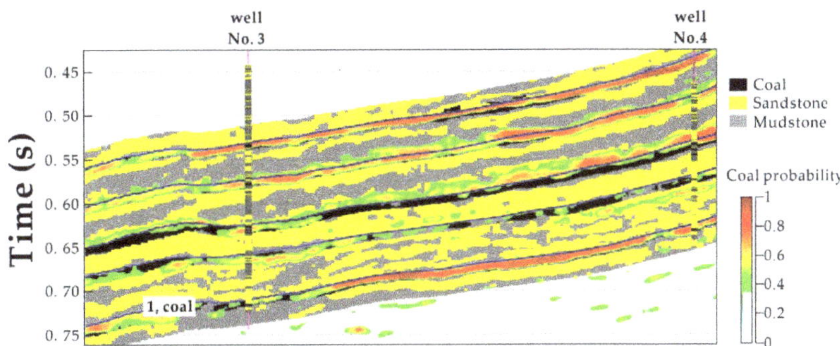

Figure 14. Inversion profile of lithology and reservoir holdup.

4. Results

(1) Reserve rock physical characteristics.

Three gas reservoirs are superimposed and developed in this area. The gas-bearing reservoir parameters interpreted by logging are different. Among them, coal bed methane reservoirs generally contain gas; gas saturation is relatively high, reaching above 50%; and the gas saturation of tight sandstone gas reservoirs and shale gas reservoirs is less than 10% (Figure 3).

After comparative analysis, it can be seen that the junction of P-wave impedance and P-wave to S-wave velocity ratio can show the petrophysical distribution of the reservoir more clearly. From this, it can be seen that the gas-bearing reservoir area seriously overlaps the tight sandstone and mudstone non-reservoir areas. Sandstone areas and mudstone non-reservoir areas can be effectively distinguished (Figure 4)

The gas-bearing reservoirs show the characteristics of $\lambda \cdot \rho$ and $\mu \cdot \rho$ double low (Figure 5).

(2) Gas-bearing distribution of coal measure strata.

Based on the analysis and interpretation of the inversion results of various reservoir parameters, the qualitative and quantitative research system is used to delineate the favorable areas and comprehensively predict the gas-bearing reservoirs. There are five sets of main coal seams in the study area. The study shows that the gas source of the Huainan coal measure strata is mainly coal bed methane. Considering that the coal bed methane has the characteristics of near-source enrichment and accumulation, the buried depth is a very important factor for coal measure strata [28,29]. So, this sweet spot prediction selected $1_\#$ coal seam as the object of research and analysis.

As mentioned earlier, the double low characteristics of $\lambda \cdot \rho$ and $\mu \cdot \rho$ can represent gas-bearing reservoirs, and the distribution of gas-bearing reservoirs in space can be explained uniformly by the method of lithological fluid probability analysis. Figure 15 shows the gas-bearing probability plan of CBM in $1_\#$ coal (red color in the figure represents high gas-bearing probability). The location of $1_\#$ coal seam in the northern part of the study area is relatively high, and the high probability of gas-bearing reservoirs are widely distributed, which is the first choice for CBM mining. In addition, there is a high probability of gas inclusion in the vicinity of well No. 6.

(3) Lithology distribution of coal measures strata.

Based on the lithology prediction, a plan of the thickness of the $1_\#$ coal seam was obtained. It can be seen from the map that the thickness of the $1_\#$ coal seam in the study area is thicker and well-developed, and the southern part is relatively wide, which is generally beneficial to the exploitation of CBM.

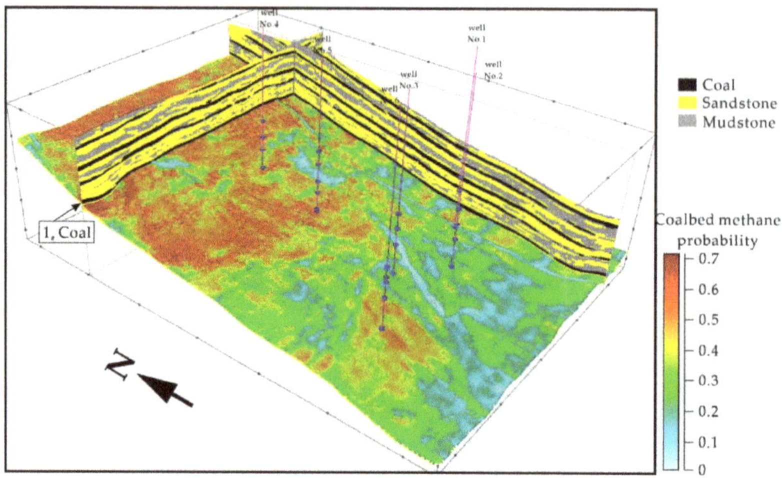

Figure 15. $1_\#$ coal seam gas reservoir distribution map.

(4)   Analysis of brittleness index of coal measure formation.

The brittleness index plays a vital role in exploiting gas-bearing reservoirs in coal-measure formations, mainly when the utilization level is used for production. It can be seen from the plan of brittleness index extracted from $1_\#$ coal seam (Figure 16) that the red area indicates that the coal-measure formation has a high brittleness index, which is mainly located in the northern part of the work area, and cracks are prone to occur during fracturing. In contrast, the brittleness index of the formation in the southern region is low, which is not easy to fracture.

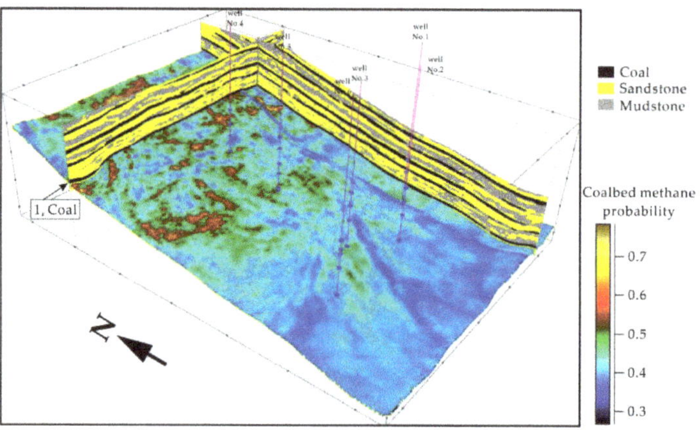

Figure 16. Plan of brittleness index of $1_\#$ coal seam.

The prediction of the excellent enrichment area of superimposed and coexisting gas-bearing reservoirs mainly refers to the three reservoir parameters of high gas-bearing capacity, thick coal, and high brittleness index, combined with the characteristics of easy preservation of coal seam gas in areas with deeper structural depths. To delineate the favorable areas of CBM enrichment in the $1_\#$ coal seam. The measured results from 72 actual test CBM extraction wells from coal mines were compared with the predicted results. The results show that the CBM reservoirs in relatively high gas content areas are thicker (>0.005 s) and brittle index (>0.5), and the gas content is also high (>0.5). Finally, the favorable CBM reservoir areas in the study area were delineated, and there were four final delineated areas (Figure 17).

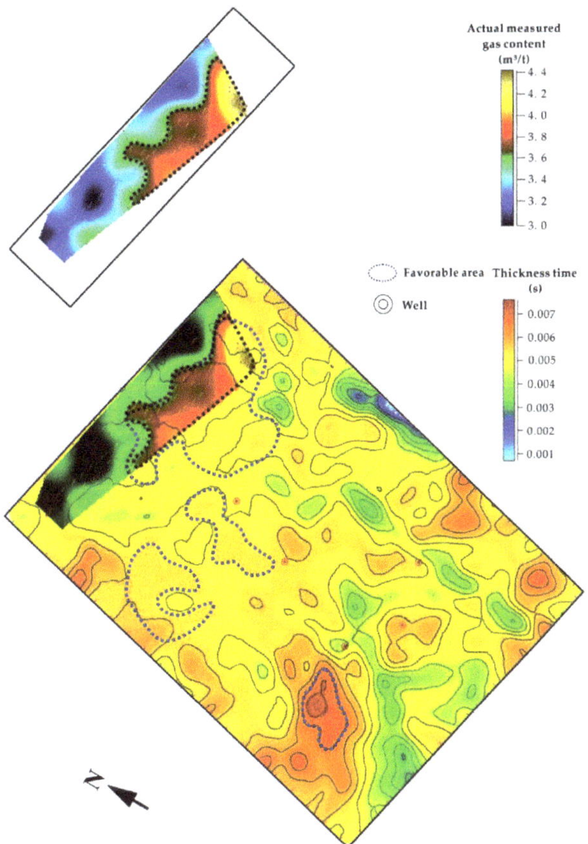

Figure 17. Comparison between actual measurement and prediction results of CBM content.

5. Discussions

The coal seams in the Zhangji coal mine area in Huainan, China are characterized by multiple overlapping layers, which makes the prediction of CBM more difficult. The drilling method has higher detection accuracy for CBM, but the economic cost is higher for large area detection. The geological survey method can realize the detection of large area at lower cost, but the detection accuracy is poor. Considering that the seismic exploration technology has been widely promoted in China's coal field, the 3D seismic prestack inversion method has been successfully used in oil and gas field for reservoir prediction. In this study, the coal seam thickness, tectonics, and petrophysical conditions are considered. Firstly, a petrophysical analysis of the physical properties of CBM reservoirs was conducted, and the correlation between the P-wave impedance of CBM reservoirs and the lithology, fluid factor $\lambda \cdot \rho$, $\mu \cdot \rho$, and brittleness index is summarized. Then, the joint response of CBM fluid factor, coal seam spreading, and brittle distribution information is combined to predict the magnitude of gas-bearing potential of CBM. In order to verify the reliability of the results, statistical analysis of the CBM extraction well data in the existing area of the site was carried out, and the CBM distribution map representing the actual coal content size was drawn and analyzed with the predicted results. The results show that using the criteria of thickness > 0.005 s, brittleness index > 50%, and gas content > 50% of the CBM reservoir, the possibility of CBM in the study area can be predicted. This method extends the application of the pre-stack seismic simultaneous inversion method in the Huainan coal region of China based on the previous work and improves the accuracy of prediction by combining coal bed distribution, brittleness index, and fluid factor to predict the gas-bearing potential of coalbed methane.

6. Conclusions

Based on the logging data for CBM reservoir petrophysical analysis, the 3D seismic prestack simultaneous inversion method and lithological fluid probability analysis are used to obtain reservoir gas content, brittleness index, reservoir lithology, and thickness information, and to achieve the prediction of the best favorable zone for CBM. The point of this method is that it integrates geological, seismic, and logging data and takes into account various characteristics of the coal thickness, fold structure, gas content, and brittleness index of the CBM reservoir, which improves the accuracy of prediction. From the petrophysical analysis results, the "double low" characteristics of $\lambda \cdot \rho$ and $\mu \cdot \rho$ of CBM reservoir can be derived. By comparing the results with the experimental wells for CBM extraction, it is found that the thick coal seam, high brittleness index, and high gas content can better reflect the best favorable zone for CBM reservoirs. Therefore, the method proposed in this study is important for future CBM and even three gas predictions, reducing the prediction cost and improving the development efficiency.

Author Contributions: B.C.: Conceptualization, Methodology, Writing—Original draft preparation; F.C.: Supervision, Writing—Review and Editing, Validation; B.L.: Resources, Validation; Y.D.: Data curation, Validation; G.D.: Data curation; C.W.: Validation; Z.W.: Validation; R.W.: Visualization. All authors have read and agreed to the published version of the manuscript.

Funding: This work was supported by the open project of State Key Laboratory of Coal Resources and Safe Mining (China University of Mining and Technology-Beijing) (SKLCRSM21KFA04).

Institutional Review Board Statement: Not applicable.

Informed Consent Statement: Not applicable.

Data Availability Statement: Not applicable.

Conflicts of Interest: The authors declare no conflict of interest.

References

1. Xinhua, M. "Extreme Utilization" Development Theory of Unconventional Natural Gas. *Pet. Explor. Dev.* **2021**, *48*, 381–394. [CrossRef]
2. Jianchao, H.; Zhiwei, W.; Pingkuo, L. Current States of Coalbed Methane and Its Sustainability Perspectives in China. *Int. J. Energy Res.* **2018**, *42*, 3454–3476. [CrossRef]
3. Tang, B.; Tang, Y.; Wang, C.; Li, Y.; Yu, Y.; Dai, M.; Wang, Z. Rapid co-extraction of coal and coalbed methane techniques: A case study in Zhangji Coal Mine, China. In *Proceedings of the 2019 5th International Conference on Advances in Energy Resources and Environment Engineering (Icaesee 2019)*; Iop Publishing Ltd.: Bristol, UK, 2020; Volume 446, p. 052012.
4. Wei, Q.; Li, X.; Hu, B.; Zhang, X.; Zhang, J.; He, Y.; Zhang, Y.; Zhu, W. Reservoir Characteristics and Coalbed Methane Resource Evaluation of Deep-Buried Coals: A Case Study of the No.13-1 Coal Seam from the Panji Deep Area in Huainan Coalfield, Southern North China. *J. Pet. Sci. Eng.* **2019**, *179*, 867–884. [CrossRef]
5. Fu, X.; Qin, Y.; Wang, G.G.X.; Rudolph, V. Evaluation of Coal Structure and Permeability with the Aid of Geophysical Logging Technology. *Fuel* **2009**, *88*, 2278–2285. [CrossRef]
6. Zhang, X.M.; Li, J.W.; Han, B.S.; Dong, M.T. Division and Formation Mechanism of Coalbed Methane Reservoir in Huainan Coalfield, Anhui Province, China. *Chin. Sci. Bull.* **2005**, *50*, 7–17. [CrossRef]
7. Zhou, L.; Li, J.; Chen, X.; Liu, X.; Chen, L. Prestack Amplitude versus Angle Inversion for Young's Modulus and Poisson's Ratio Based on the Exact Zoeppritz Equations. *Geophys. Prospect.* **2017**, *65*, 1462–1476. [CrossRef]
8. Zhang, F.; Yang, J.; Li, C.; Li, D.; Gao, Y. Direct Inversion for Reservoir Parameters from Prestack Seismic Data. *J. Geophys. Eng.* **2020**, *17*, 993–1004. [CrossRef]
9. Kang, Y.; Shang, C.; Zhou, H.; Huang, Y.; Zhao, Q.; Deng, Z.; Wang, H.; Ma, Y.Z. Mineralogical Brittleness Index as a Function of Weighting Brittle Minerals-from Laboratory Tests to Case Study. *J. Nat. Gas Sci. Eng.* **2020**, *77*, 103278. [CrossRef]

10. Russell, B.H.; Hedlin, K.; Hilterman, F.J.; Lines, L.R. Fluid-property Discrimination with AVO: A Biot-Gassmann Perspective. *Geophysics* **2003**, *68*, 29–39. [CrossRef]
11. Yasin, Q.; Sohail, G.M.; Liu, K.-Y.; Du, Q.-Z.; Boateng, C.D. Study on Brittleness Templates for Shale Gas Reservoirs—A Case Study of Longmaxi Shale in Sichuan Basin, Southern China. *Pet. Sci.* **2021**, *18*, 1370–1389. [CrossRef]
12. Gullapalli, S.; Dewangan, P.; Kumar, A.; Dakara, G.; Mishra, C.K. Seismic Evidence of Free Gas Migration through the Gas Hydrate Stability Zone (GHSZ) and Active Methane Seep in Krishna-Godavari Offshore Basin. *Mar. Pet. Geol.* **2019**, *110*, 695–705. [CrossRef]
13. Lu, J.; Wang, Y.; Chen, J.; An, Y. Joint Anisotropic Amplitude Variation with Offset Inversion of PP and PS Seismic Data. *Geophysics* **2018**, *83*, N31–N50. [CrossRef]
14. Tan, P.; Jin, Y.; Han, K.; Zheng, X.; Hou, B.; Gao, J.; Chen, M.; Zhang, Y. Vertical Propagation Behavior of Hydraulic Fractures in Coal Measure Strata Based on True Triaxial Experiment. *J. Pet. Sci. Eng.* **2017**, *158*, 398–407. [CrossRef]
15. Wang, P.; Chen, X.; Wang, B.; Li, J.; Dai, H. An Improved Method for Lithology Identification Based on a Hidden Markov Model and Random Forests. *Geophysics* **2020**, *85*, IM27–IM36. [CrossRef]
16. Gale, J.F.W.; Reed, R.M.; Holder, J. Natural Fractures in the Barnett Shale and Their Importance for Hydraulic Fracture Treatments. *AAPG Bull.* **2007**, *91*, 603–622. [CrossRef]
17. Huo, Z.; Zhang, J.; Li, P.; Tang, X.; Yang, X.; Qiu, Q.; Dong, Z.; Li, Z. An Improved Evaluation Method for the Brittleness Index of Shale and Its Application—A Case Study from the Southern North China Basin. *J. Nat. Gas Sci. Eng.* **2018**, *59*, 47–55. [CrossRef]
18. Rickman, R.; Mullen, M.; Petre, E.; Grieser, B.; Kundert, D. A Practical use of shale petrophysics for stimulation design optimization: All shale plays are not clones of the barnett shale. In Proceedings of the SPE Annual Technical Conference and Exhibition, Denver, CO, USA, 21–24 September 2008; OnePetro: Richardson, TX, USA, 2008; pp. 1–11.
19. Chen, X.-P.; Huo, Q.; Lin, J.; Wang, Y.; Sun, F.; Li, W.; Li, G. Theory of CBM AVO: I. Characteristics of Anomaly and Why It Is So. *Geophysics* **2014**, *79*, D55–D65. [CrossRef]
20. Peng, S.; Chen, H.; Yang, R.; Gao, Y.; Chen, X. Factors Facilitating or Limiting the Use of AVO for Coal-Bed Methane. *Geophysics* **2006**, *71*, C49–C56. [CrossRef]
21. Wang, Y.; Cui, R.; Zhang, S.; Dai, F.; Jia, W. Application of AVO Inversion to the Forecast of Coalbed Methane Area. In *Proceedings of the 2011 Xi'an International Conference on Fine Exploration and Control of Water & Gas in Coal Mines*; Shuning, D., Qun, Z., Li, W., Yongsheng, Q., Yujing, X., Hong, F., Eds.; Elsevier Science: Amsterdam, The Netherlands, 2011; Volume 3, pp. 210–216.
22. Yan, X.; Zhu, Z.; Hu, C.; Gong, W.; Wu, Q. Spark-Based Intelligent Parameter Inversion Method for Prestack Seismic Data. *Neural Comput. Appl.* **2019**, *31*, 4577–4593. [CrossRef]
23. Nolet, G. *Quantitative Seismology, Theory and Methods*; Elsevier: Amsterdam, The Netherlands, 1980.
24. Wu, Q.; Zhu, Z.; Yan, X. Research on the Parameter Inversion Problem of Prestack Seismic Data Based on Improved Differential Evolution Algorithm. *Clust. Comput. J. Netw. Softw. Tools Appl.* **2017**, *20*, 2881–2890. [CrossRef]
25. Yenwongfai, H.D.; Mondol, N.H.; Faleide, J.I.; Lecomte, I. Prestack Simultaneous Inversion to Predict Lithology and Pore Fluid in the Realgrunnen Subgroup of the Goliat Field, Southwestern Barents Sea. *Interpret. J. Subsurf. Charact.* **2017**, *5*, SE75–SE96. [CrossRef]
26. Zhao, L.; Geng, J.; Cheng, J.; Han, D.; Guo, T. Probabilistic Lithofacies Prediction from Prestack Seismic Data in a Heterogeneous Carbonate Reservoir. *Geophysics* **2014**, *79*, M25–M34. [CrossRef]
27. Buland, A.; Kolbjornsen, O.; Hauge, R.; Skjaeveland, O.; Duffaut, K. Bayesian Lithology and Fluid Prediction from Seismic Prestack Data. *Geophysics* **2008**, *73*, C13–C21. [CrossRef]
28. Wei, Q.; Hu, B.; Li, X.; Feng, S.; Xu, H.; Zheng, K.; Liu, H. Implications of Geological Conditions on Gas Content and Geochemistry of Deep Coalbed Methane Reservoirs from the Panji Deep Area in the Huainan Coalfield, China. *J. Nat. Gas Sci. Eng.* **2021**, *85*, 103712. [CrossRef]
29. Zhang, Z.; Qin, Y.; You, Z.; Yang, Z. Distribution Characteristics of In Situ Stress Field and Vertical Development Unit Division of CBM in Western Guizhou, China. *Nat. Resour. Res.* **2021**, *30*, 3659–3671. [CrossRef]

Article

Analysis of Available Conditions for InSAR Surface Deformation Monitoring in CCS Projects

Tian Zhang [1,2], Wanchang Zhang [1,*], Ruizhao Yang [3], Huiran Gao [4] and Dan Cao [1,2]

[1] Key Laboratory of Digital Earth Science, Aerospace Information Research Institute, Chinese Academy of Sciences, Beijing 100094, China; zhangtian@radi.ac.cn (T.Z.); caodan@radi.ac.cn (D.C.)
[2] University of Chinese Academy of Sciences, Beijing 100049, China
[3] School of Geosciences & Surveying Engineering, China University of Mining and Technology-Beijing, Beijing 100083, China; yrz@cumtb.edu.cn
[4] National Institute of Natural Hazards, Ministry of Emergency Management of China, Beijing 100085, China; gaohr@radi.ac.cn
* Correspondence: zhangwc@radi.ac.cn; Tel.: +86-10-8217-8131

Abstract: Carbon neutrality is a goal the world is striving to achieve in the context of global warming. Carbon capture and storage (CCS) has received extensive attention as an effective method to reduce carbon dioxide (CO_2) in the atmosphere. What follows is the migration pathway and leakage monitoring after CO_2 injection. Interferometric synthetic aperture radar (InSAR) technology, with its advantages of extensive coverage in surface deformation monitoring and all-weather traceability of the injection processes, has become one of the promising technologies frequently adopted in worldwide CCS projects. However, there is no mature evaluation system to determine whether InSAR technology is suitable for each CO_2 sequestration area. In this study, a new evaluation model is proposed based on the eight factors that are selected from the principle of the InSAR technique and the unique characteristics of the CO_2 sequestration area. According to the proposed model, the feasibility of InSAR monitoring is evaluated for the existing typical sequestration areas in the world. Finally, the challenges and prospects of InSAR in the CCS project are discussed.

Keywords: InSAR monitoring; carbon capture and storage; feasibility assessment model

Citation: Zhang, T.; Zhang, W.; Yang, R.; Gao, H.; Cao, D. Analysis of Available Conditions for InSAR Surface Deformation Monitoring in CCS Projects. *Energies* **2022**, *15*, 672. https://doi.org/10.3390/en15020672

Academic Editors: Sohrab Zendehboudi and Muhammad Aziz

Received: 2 November 2021
Accepted: 16 January 2022
Published: 17 January 2022

Publisher's Note: MDPI stays neutral with regard to jurisdictional claims in published maps and institutional affiliations.

1. Introduction

The constant excess of carbon dioxide (CO_2) in the atmosphere is an important issue that human beings need to solve now and in the future. The Global Carbon Project (GCP) has published its latest assessment of the global CO_2 budget, and despite the impact of COVID-19, human-made CO_2 emissions (CO_2 emissions from fossil fuel consumption and land use change) are still far greater than the net CO_2 absorption by oceans and lands [1]. CO_2 sequestration and utilization programs remain one of the most effective and cost-effective methods available [2]. With the development of sequestration technology and an increase in sequestration sites, it is urgent and necessary to ensure the safety and effective monitoring of sequestration.

CO_2 will exist in a gas or supercritical fluid state after being injected into the reservoir, and the density of injected CO_2 is usually lower than that of resident fluid. Due to gravity or the buoyancy effect, it may flow upward and leak to the ground; however, the density of CO_2 is greater than that of air. After leakage, it will accumulate on the surface, which may lead to the loss of humans, animals, plants, and environment [3]. Therefore, real-time ground monitoring during injection is necessary. At the same time, due to the change of formation pressure caused by injection, new fractures may be generated, which will cause CO_2 not to migrate according to the expected pathway but integrate into groundwater or rise to the surface along the fractures, resulting in the failure of storage, and the expected effect cannot be achieved in gas or oil displacement projects. Therefore, large-scale and long-time migration monitoring is also needed [4]. In order to realize real-time and long-term monitoring at the same time, multiple monitoring methods need to be applied jointly, which also leads to a large amount of investment and an increase in monitoring cost. A systematic and economic monitoring system still needs to be developed [5].

Currently, the practices utilized in leakage monitoring methods are primarily divided into three categories: near-surface monitoring, surface monitoring, and underground monitoring [6]. Among them, interferometric synthetic aperture radar (InSAR) monitoring, as an emerging near-surface monitoring technology, has attracted extensive attention, and its effect and feasibility for storage monitoring have been verified [7]. The successful monitoring cases of satellite-borne synthetic aperture radar (SAR) show that InSAR monitoring technology can conduct large-scale all-day monitoring without the need for a large number of personnel and instruments on site. Additionally, InSAR technology can realize continuous monitoring for years, and can effectively identify the underground migration direction of CO_2, which is a reliable monitoring method from the perspectives of the economy and sustainability [8].

However, the use of InSAR technology in CCS is still in the exploratory stage. Although a lot of funds have been invested, the monitoring performance of InSAR in some sequestration areas is not ideal, which is primarily reflected in problems with identification and accuracy. InSAR technology still has great challenges in CCS project monitoring, such as incoherence caused by atmosphere and vegetation, which will lead to failure to detect deformation [9,10]. As a passive detection method, how to separate the deformation caused by CO_2 injection is also a problem to be solved [11]. These problems lead to different effects in different projects. Thus, a systematic analysis and summary is urgently needed. It not only provides experience for existing projects, but also provides the basis for the evaluation of InSAR feasibility for subsequent storage monitoring projects.

In this study, using an analysis of successful monitoring cases, a feasibility evaluation model of InSAR surface deformation monitoring in CCS project was established. Eight factors, including the limitations of InSAR technology and the characteristics of storage area, were selected as the main controlling factors. Then, the model was applied to evaluate storage areas all over the world and give suggestions on whether it is suitable for InSAR monitoring. The study is structured as follows. After the introduction, the principle of deformation induced by CO_2 injection and the feasibility of InSAR monitoring will be discussed in Section 2. Then, factors that may affect InSAR sequestration monitoring are proposed in Section 3, followed by a feasibility assessment of an existing/in design sequestration area in Section 4. Finally, a summary and discussion regarding the direction of improvements of the feasibility assessment and the future development of InSAR technology are discussed.

2. Feasibility of Monitoring CCS Using the InSAR Technique

During the storage process of CCS, when CO_2 is injected, the pore pressure in the rock layer increases, which changes the effective stress field. This then causes a deformation of the reservoir, resulting in vertical strain and damage. In particular, the deformation near the injection point is the most sensitive and obvious, and this phenomenon has been confirmed in a variety of numerical simulation models [12–14].

In terms of its primary "measurement" function, InSAR is a space–earth observation technology that can measure millimeter-scale ground movement. In terms of its working principle, it realizes the measurement of spatial geometry and its changes by "comparing" the "phase change" (echo path length) in two or more radar image data of the same area [15]. The millimeter–centimeter deformation measurement accuracy of the InSAR technique is particularly consistent with the centimeter deformation measurement in the CO_2 sequestered area. Therefore, the corresponding research began shortly after the InSAR technique was proposed. In 2004, InSAR monitoring was first used in the In Salah field to effectively monitor the 5–150 mm/y surface deformation caused by CO_2 injection [16]. With the continuous development and improvement of InSAR technology, more sensors and a time series unwrapping algorithm have been used to monitor the CCS deformation. These include the application of an L-band sensor in the Scurry Country CO_2-EOR field in West Texas [17], the application of a C-band sensor in the Jingbian CO_2-EOR field in Shaanxi [18], China, the application of D-InSAR technology in the Aquistore CCS site in southeastern Saskatchewan, Canada [19], and the application of SBAS-InSAR technology in the Fengcheng Oil Field, Xinjiang, China [20].

Currently, more than 137 CCS projects have been built or planned in the world, as shown in Figure 1, while less than 10 cases have been successfully monitored using InSAR technology. As a large-scale, long-term, and economic technology, InSAR is worthy of more promotion, but the uncertainty of its monitoring result success has prevented it from being widely used. In order to avoid the wastage of funds caused by the failure of application, a method to evaluate the feasibility of InSAR in the storage area is urgently required.

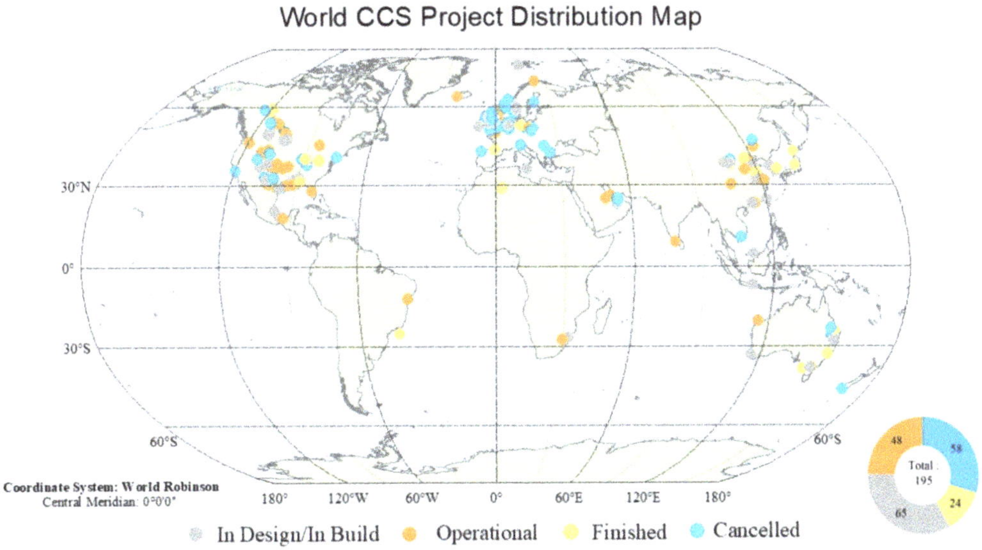

Figure 1. CCS project distribution map.

3. Influencing Factors of InSAR Monitoring in a CCS Project

Based on the principle of InSAR and a review of the successful cases, this study proposes eight factors that may affect CCS monitoring using the InSAR technique. These can be divided into two factors of InSAR's own limitations and six factors specific to the CCS storage area.

3.1. Vegetation Coverage

Phase unwrapping is the process of restoring phase from the principal value/phase to the true value, which is also a core step of the InSAR technique. The accuracy of phase unwrapping directly affects the final surface deformation results [21]. In 1989, when the California Institute of Technology demonstrated the feasibility of InSAR technology for detecting surface deformation for the first time, it was found that there was an incoherent phenomenon in the vegetated areas that led to a failure of the phase unwrapping [22]. The linear relationship between the normalized difference vegetation index (NDVI) and the SAR interferometric coherence has also been revealed [23]. Following years of exploration, time series InSAR was proposed to effectively reduce the incoherent effects caused by vegetation, but it is still deficient in dense areas.

In this study, the fractional vegetation cover (FVC) was used to measure the status of the surface vegetation. The FVC index refers to the percentage of the vertical projection area of vegetation on the ground to the total area of the statistical area [24]. The practical method to measure the FVC index is to calculate it using the NDVI, and the calculation equation is as follows:

$$FVC = (NDVI - NDVI_{soil})/(NDVI_{veg} - NDVI_{soil}), \tag{1}$$

where $NDVI_{soil}$ refers to the NDVI value of bare soil or areas without vegetation coverage; and $NDVI_{veg}$ refers to the NDVI value of pixels completely covered by vegetation. In this study, according to the different regions where the carbon dioxide sequestration areas were located, the factor scores were formulated according to the range of vegetation coverage, as presented in Table 1.

Table 1. The assessment rules for the fractional vegetation cover (FVC).

FVC	Surface Type	S_{FVC} Score
0–0.1	Barren	1.0
0.1–0.3	Low coverage	0.8
0.3–0.45	Medium-low coverage	0.6
0.45–0.6	Medium coverage	0.4
0.6–1	High coverage	0.2

The Glass-FVC dataset of the University of Maryland [25] was used to extract the FVC in the 5 km rectangular buffer zone with the injection point as the center; the number of all types of grids was counted, and this was multiplied by the corresponding scoring factor. Then the average was calculated, which can be regarded as the final FVC score of the storage area. In order to intuitively illustrate our calculation method, three typical sequestration areas were selected for demonstration, as shown in Figure 2. In Salah oilfield stores CO_2 in its deep saline layer, with an average annual storage capacity of 3 Mt. Jingbian oilfield is the first CCUS full process project in China, with a storage depth of 1500 m and an average storage capacity of 59,000 tons. The Ketzin pilot is the first organization in the world with a relatively perfect monitoring system for CO_2 geological storage. The reservoir is a saltwater aquifer with a depth of about 650 m, with an average annual storage of 60,000 tons. Among them, the In Salah sequestration area is located in the Sahara Desert with no vegetation coverage and the proportion of bare land is 100%. Hence, its S_{FVC} score is one. The Jingbian oil field is located in western China and is dominated by low vegetation coverage with an FVC less than 0.3. The final S_{FVC} score was 0.794 after multiplying the proportion of the three surface types by the corresponding S_{FVC} score. In contrast, Germany's Ketzin sequestration area is more complex, with a large proportion of medium and high vegetation coverage leading to an S_{FVC} score as low as 0.409, which affects the applicability of InSAR monitoring.

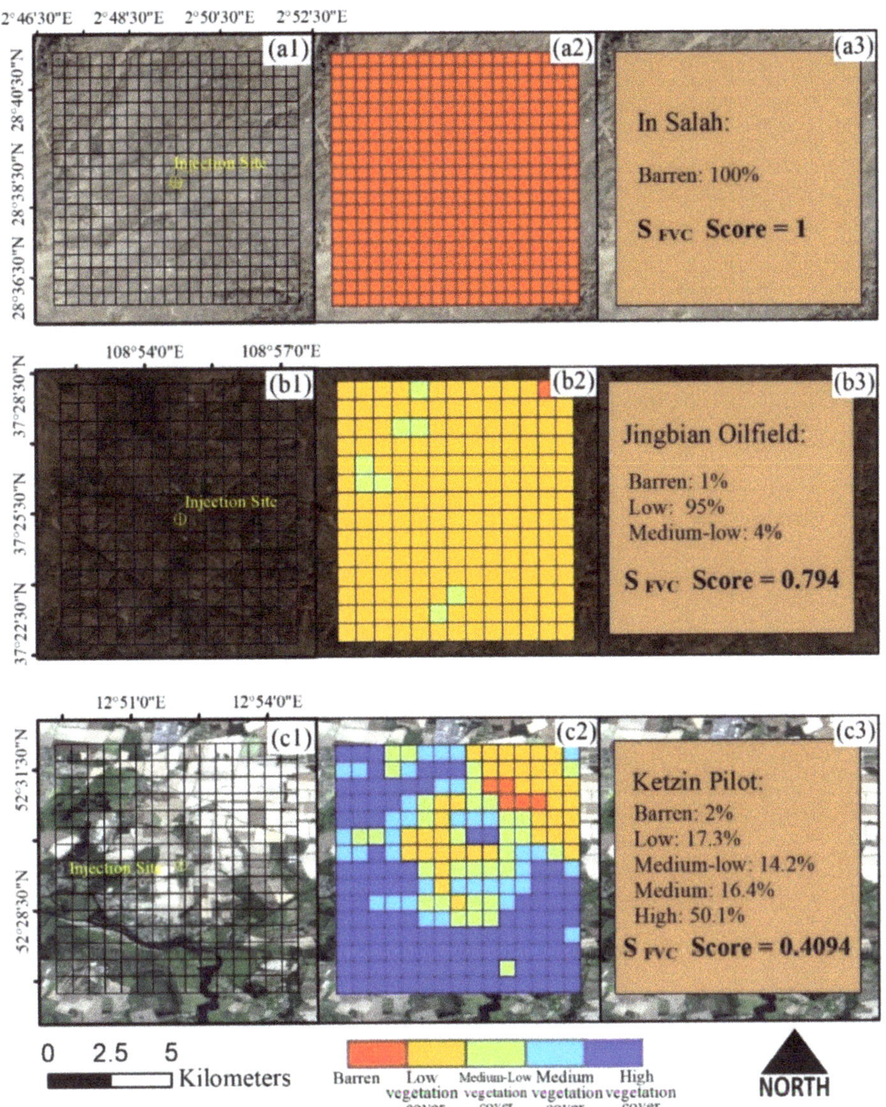

Figure 2. The graphs show the vegetation coverage of Salah, Jingbian, and Ketzin. The base map is the true color composite image of LANDSAT 8 OLI. (**a1–c1**) show the location of the injection wells and their buffers in the study area. (**a2–c2**) show their corresponding FVC index with a resolution of 500 m. (**a3–c3**) illustrate the proportion of the vegetation coverage classification in each study area, and their S_{FVC} scores.

3.2. Topographic Factor

The imaging principle of SAR requires that the image be expressed on an oblique anomaly surface. Therefore, in SAR images, when the slope of ground surface scene is too large, different ground target points with the same oblique distance will be mapped into the same range-doppler unit, and their reflected echoes will overlap. This will result in an aliasing of the corresponding SAR image and the InSAR interference phase, which is called the layout [26]. Steep terrains will also block the back slope area, leading to some object points being shaded. Hence, the radar cannot receive the corresponding surface echo information, resulting in dark areas in the SAR image, and this phenomenon is called the shadow [27]. The presence of overlay and shadow will cause the area to be incorrectly unwrapped.

The R-index, which indicates the ratio between the pixel size in slant and the ground range geometry or the "pixel compression factor", was proposed in 2011 [28]. The R-index quantifies the effect of terrain on radar imaging, and its calculation equation is as follows:

$$R_{Index} = -\sin(\arctan(\tan S \cdot \sin A\alpha) - \theta) \tag{2}$$

where S is the slope derived from the digital elevation model (DEM), $A\alpha$ is the aspect derived from DEM and correct with angle from north of the satellite track, and θ is the incident angle of the line of sight (LOS). The higher the *R-index*, the flatter the terrain, with less shadow and layout. The factor scores corresponding to the R-index are listed in Table 2. The score of the topographic factors in a sequestration area was calculated by adding the percentage of each R class multiplied by the corresponding scoring factors.

Table 2. The assessment rules of the R_{Index}.

R_{Index} Classes	Pixel Compression	$S_{Terrain}$ Score
≤ 0	Layout/foreshortening	0
0–0.3	High–bad slope	0.25
0.3–0.6	Medium slope	0.5
0.6–0.8	Low good slope	0.75
>0.8	Very low good slope	1

The CCS-ECBM pilot of Shizhuang, Shaanxi Province, China was selected to illustrate the calculation method of the $S_{terrain}$ score. Shizhuang pilot data were DEM generated from unmanned aerial vehicle (UAV) images, and the slope and aspect parameters were extracted from the DEM to calculate the corresponding R_{Index}. The raster proportion of the R_{Index} was multiplied by the corresponding $S_{terrain}$ score to finally obtain the $S_{terrain}$ score of the sequestration area, as shown in Figure 3.

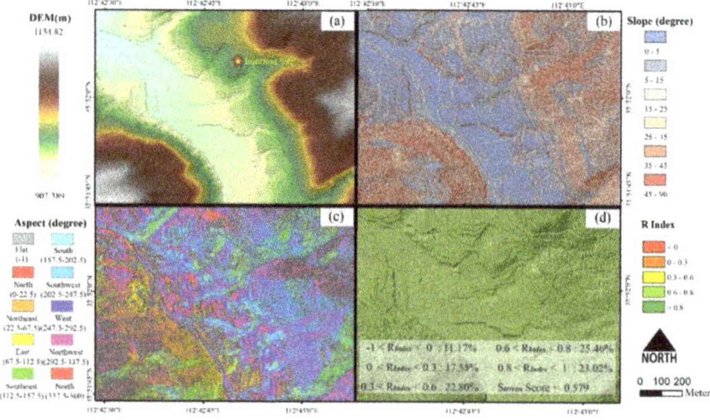

Figure 3. The graphs show the topographic parameters of Shizhuang, Shaanxi Province, China. (**a**) shows the DEM extracted using a UAV. (**b**) and (**c**) show the slope and aspect parameters extracted from the DEM. (**d**) illustrates the R_{Index} of the injection area and the $S_{terrain}$ score calculated from the distribution.

3.3. Reservoir Location

Currently, CCS sequestration forms are primarily divided into geological sequestration and marine sequestration, among which geological storage is the mainstream, while marine storage has more application potential. Geological sequestration techniques can be roughly divided into three categories: (1) oil and gas reservoir sequestration [29], (2) unrecoverable coal seam sequestration [30], and (3) deep saltwater sequestration [31]. All of these sequestration forms will cause changes in the surface elevation due to changes in the underground pressure. Marine sequestration is the direct injection of CO_2 into the ocean at high pressure after CO_2 capture, which can be divided into: (1) shallow sea dissolution sequestration (200–300 m), (2) deep-sea clathrate sequestration (>500 m), and (3) deep-sea clathrate hydrate sequestration (>3000 m) [32]. After centuries of atmospheric isolation, dissolved and dissipated CO_2 becomes part of the global carbon cycle and is an ideal way to potentially achieve large-scale and long-term CO_2 sequestration [33].

Obviously, with the InSAR technique, there is no way to observe marine sequestration. Therefore, we utilized the sequestration form as a prerequisite for the application of the InSAR technology. The onshore sequestration score was one, and the offshore sequestration score was zero, which was multiplied with all the other factor scores as a coefficient, as shown in Table 3.

Table 3. The assessment rules of the reservoir location.

Reservoir Location	$S_{location}$ Score
Onshore	1
Offshore	0

3.4. Land Use/Land Cover

InSAR technology is unable to detect large magnitudes of deformation or large changes in the surface due to incoherent reasons, such as human activity images in intense areas. Therefore, distinguishing the surface environment of different sequestration areas is an important factor for the InSAR technique.

The land use and land cover (LULC) is an important indicator for the effective observation of global surface types [34]. The 30 m resolution LULC data published by the National Remote Sensing Center of China (NRSCC) in 2017 have been validated and widely used all over the world [35]. Among the 10 categories defined by the NRSCC, there are 5 categories of natural vegetation, 2 categories of land use and land mosaic, and 3 categories of land without plant growth [36]. In InSAR monitoring, the overall situation of the surface is only required. Hence, we combined similar categories. Bare land is an ideal site for monitoring InSAR technology due to its good coherence caused by an abundance of bare land surfaces [37]. Since InSAR can be monitored in a long time series according to a fixed season, monitoring can be conducted when seasonal plants wilt to ensure coherence, which makes the location of seasonal vegetation an appropriate location for InSAR monitoring [38]. With regard to urban areas and croplands, anthropogenic activities greatly disturb the surface change, so it is also necessary to conduct an analysis or experiment according to the specific site situation. However, for water bodies and wetlands, there are too many factors that affect their changes, which is why InSAR technology is not applicable to monitor elevation changes caused by CO_2 injection. The specific factor scores of the classifications are presented in Table 4.

The Scurry and San Juan oilfields were used to introduce the calculation of the S_{lulc} score. Similarly, a rectangular buffer of 10 km at the injection point was selected as the calculation area. The total S_{lulc} score of the sequestration area was obtained by calculating the proportion of the different types of grids and multiplying by their S_{lulc} score, as shown in Figure 4. The Scurry storage area consisted primarily of grassland and shrubland, which accounted for 76% of the buffer, and the S_{lulc} score was 0.728, which may have different effects on InSAR monitoring with seasonal changes. The San Juan oilfield was primarily bare land, accounting for 83% of the buffer. Hence, its S_{lulc} score reached 0.951, which was more suitable for InSAR monitoring from the perspective of the LULC.

Table 4. The assessment rules of the land use/land cover (LULC).

Applicability of InSAR Technique	NRSCC Classification	S_{lulc} Score
Very suitable	Bare land	1.0
Suitable	Grassland Impervious surface	0.8
Suspectable/As appropriate	Forest Tundra Cropland Shrubland	0.5
Not suitable	Water Wetland Snow and Ice	0

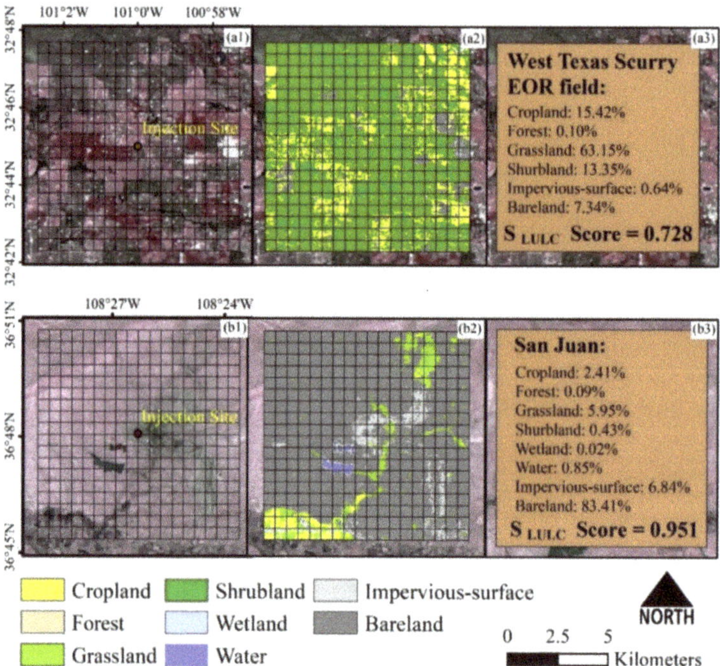

Figure 4. The graphs show the LULC of Scurry and San Juan. The base map is the true color composite image of LANDSAT 8 OLI. (**a1,b1**) show the location of the injection wells and their buffers in the study area. (**a2,b2**) show their LULC classification with a resolution of 30 m. (**a3,b3**) illustrate the proportion of the LULC classification in each study area, and the calculated S_{LULC} score.

3.5. Injection Rate

The injection rate is the primary factor that affects the formation pressure and stress. In a low-velocity injection, the fluid can move through the formation at a lower differential pressure and the formation pressure rises slowly. With an increase in the injection rate, it becomes more difficult for the fluid to flow and diffuse in the formation, and the required driving pressure increases [39]. According to the numerical simulation, a good linear relationship between the surface deformation and injection velocity was obtained when the injection rate was low (30–150 m^3/day), while a quadratic relationship was obtained when the injection rate was high (>2000 m^3/day) [40].

The obvious deformation is very beneficial for InSAR monitoring, and the factor scores are listed in Table 5. However, it should be noted that if the vertical displacement rate of the surface is too large, the fracture expansion of the stratum will increase, and the probability of fault activity will increase. This will lead to geological disasters on the surface and damage to the ecological environment on the surface. Therefore, it is necessary to control the injection rate according to the bottom hole pressure to ensure that the surface deformation rate is within a reasonable range.

Table 5. The assessment rules of the injection rate.

Injection Rate (m^3/Day)	S_{Rate} Score
<30	0.2
30–150	0.4
150–2000	0.6
2000–8000	0.8
>8000	1

3.6. Injection Quantity

During the process of CO_2 sequestration, the matrix injection (injection pressure lower than the fracture pressure of the reservoir) is used to prevent leakage caused by the fracture of the cap rock. Matrix injection causes the reservoir pressure to build up, reducing the effective stress and causing reservoir swelling. This, in turn, lifts the cap rock and the ground surface. The total injection quantity is one of the important factors that determine the internal pressure of formation. A small amount of CO_2 injection can result in surface changes that are not obvious or concentrated near the injection well, making it difficult to be monitored using the InSAR technique. Therefore, factor scores have been formulated according to the amount of storage, as presented in Table 6. The effect of a small amount of CO_2 injection is not obvious, but it does not mean that it is completely unfeasible, and it needs to be determined in conjunction with the depth and lithology of the formation. Hence, the minimum factor score utilized in this study is not zero.

Table 6. The assessment rules of the injection quantity.

Total Injection Quantity (Mt)	$S_{Quantity}$ Score
<0.01	0.2
0.01–0.1	0.4
0.01–0.05	0.6
0.05–1.0	0.8
>1.0	1

3.7. Reservoir Depth

Surface deformation caused by carbon dioxide injection needs to be gradually transferred from the injection layer to the surface; so, the depth of injection is also an important factor for the success of InSAR monitoring [41]. Theoretically, the shallower the storage is, the better the surface deformation effect. However, the minimum depth for supercritical CO_2 is 800 m, and the cap rock thickness is also required; hence, most of the storage thickness is between 0.8 and 5 km at present [42]. However, the factors that determine the sealing depth are the porosity, saturation, and other physical properties of the reservoir and also the integrity and thickness of the cap rock. Hence, the selection of the reservoir depth is restrictive. Based on an analysis of successful InSAR monitoring cases and expert experience, a relatively universal classification method is proposed, as presented in Table 7. The 0.5–1.5 km depth reservoir is primarily the coal field depth. With shallow strata, a small overburden pressure, and a fast deformation response, it is easy to realize InSAR observations. Oil and gas fields at depths of 1.5–2.5 km are the areas where commercial injection is concentrated. During the production process, the change in the formation pressure makes the surface deformation obvious, which can be clearly observed using InSAR technology. When the overburden pressure on the surface is greater than 2.5 km, it is no longer easy to cause surface deformation, so the fraction decreases with depth.

Table 7. The assessment rules of the reservoir depth.

Reservoir Depth (km)	S_{Depth} Score
0.5–1.5	1
1.5–2.5	0.8
2.5–4	0.6
>4	0.4

3.8. Monitoring Duration

Surface deformation changes caused by CO_2 injection require a certain amount of accumulation and response time, while the deformation response caused by fluid migration is even more delayed. Therefore, InSAR monitoring needs to be observed continuously for several years from the beginning of injection [43]. According to the review of the cases successfully monitored using InSAR, it was found that the surface deformation showed obvious changes between one and two years after injection. However, a phenomenon that cannot be ignored is that, after the injection was stopped, the underground pressure and stress were rebalanced, resulting in a decline in the surface deformation. The stratum subsidence lasted primarily for half a year after the injection was stopped, and the rate of subsidence after half a year was nearly negligible. With an increase in the injection years, the surface deformation will accumulate, and the recovery rate of the surface deformation will also decrease. According to Zheng's study, the surface recovery rate was 35.3% one year after injection, and only 6.7% when the injection period was five years [40].

Based on the above reasons, we developed factor scores for different durations, as presented in Table 8. We assumed that the InSAR monitoring began with the commencement of injection. A duration of 3–5 years is ideal for InSAR monitoring because the effects of CO_2 sequestration and migration on the surface deformation can be identified using accumulation and comparison over a period of time. When the duration is less than three years, the effect of the injection stoppage may be superimposed on the effect of the migration, making it difficult to identify and interpret. Conversely, a duration of more than five years of CO_2 injection may no longer be the primary cause of surface change and may also cause excessive expenditure, so its factor score was also lowered.

Table 8. The assessment rules of the monitoring duration.

Duration (Year)	$S_{Duration}$ Score
<1	0.5
1–3	0.8
3–5	1.0
>5	0.8

4. InSAR Feasibility Assessment of a CCS Site

4.1. Evaluation Methods

With the gradual development of InSAR technology, its application fields have been continuously expanded, and the successful application in storage deformation monitoring has made it possible for long-term and continuous migration and leakage monitoring. However, there is no evidence that InSAR monitoring can be applied to all sequestration areas. According to the properties of InSAR technology and the characteristics of sequestration areas, a feasibility assessment method based on remote sensing was proposed for monitoring surface deformation in CO_2 sequestration areas. Eight factors were selected as evaluation factors, and independent evaluation criteria and conversion scores are provided in Section 3, according to their characteristics. The final feasibility score of the study area was obtained after the converted scores were calculated in the way of equal weight. The calculation equation is as follows:

$$F_{Insar} = S_{location} \frac{\left(\sum_{i=1}^{m} S_i + \sum_{j=1}^{n} S_j\right)}{m + n}, \tag{3}$$

where F_{Insar} is the final feasibility assessment score for the sequestration area; $S_{location}$ is the sequestration type; S_i is the characteristic factor score of the InSAR technology; S_j is the characteristic factor score of CO_2 sequestration areas; and m and n are the numbers of two factors. m and n can be increased or decreased depending on the data acquisition capacity of the different storage areas. In this study, all the factors were collected; so, the corresponding equation is as follows:

$$\sum_{i=1}^{m} S_i = S_{FVC} + S_{Terrain}, \tag{4}$$

$$\sum_{j=1}^{n} S_j = S_{LULC} + S_{Rate} + S_{Quantity} + S_{Depth} + S_{Duration}, \quad n = 5, \tag{5}$$

According to the existing successful and unsatisfactory cases and combined with relevant experience, we established the corresponding classification of the feasibility score, as presented in Table 9.

Table 9. The assessment rules of feasibility.

F_{Insar} **Score**	**Feasibility**
0	Not applicable
0.1–0.3	May be applicable
0.3–0.6	Applicable
0.6–0.9	Strongly applicable
0.9–1.0	Highly recommended

4.2. Feasibility Assessment

According to the above standards, we selected 22 representative research areas all over the world as the primary evaluation objects to evaluate the feasibility of the InSAR technology, including the areas under injection and areas where injection has been completed. Sequestration areas covering as many countries as possible were evaluated in order to provide recommendations on the feasibility of InSAR monitoring. The parameters and scores of each sequestration area are presented in Table 10.

It can be seen from the evaluation that most of the sequestration areas are suitable for InSAR monitoring. Desert areas are especially advantageous for InSAR monitoring because of their flat terrain and low vegetation coverage. Based on the cases that were collected and that have been monitored using InSAR technology, the sequestration areas that have been successfully monitored were all evaluated as highly recommended or strongly applicable. In particular, a case in which InSAR did not effectively detect deformation was evaluated as applicable, which confirmed the feasibility of this evaluation method to a certain extent.

Table 10. Typical global CCS area parameters and feasibility assessment.

Project Name	Reservoir Location	Vegetation Coverage	Topographic Factor	LULC	Injection Rate (m³/Day)	Total Injection (Mt)	Reservoir Depth (km)	Duration (Year)	Country	Scale	Status	F_{Insar}
Al Reyadah	Onshore (54.47, 24.32)	Medium-low coverage	Very low good slope	Bare land	1,170,000	1.6	2.5	2	United Arab Emirates	Commercial	Operational	0.912
Frio	Onshore (−94.80, 30.00)	Medium-low coverage	Very low good slope	Forest	87,200	1.4	1.5	4	USA	Commercial	Finished	0.881
Shizhuang ECBM Pilot	Onshore (112.71, 35.87)	Low coverage	Medium slope	Cropland	5450	0.025	0.7	2	China	Pilot	Operational	0.718
In Salah	Onshore (2.82, 28.64)	Barren	Very low good slope	Bare land	1,792,000	14.0	1.8	7	Algeria	Commercial	Finished	0.993
Ketzin	Onshore (12.87, 52.49)	High coverage	Low good slope	Cropland	54,500	0.06	0.7	3	Germany	Pilot	Finished	0.756
Lanxes Newcastle	Onshore (29.97, −27.78)	Barren	Very low good slope	Bare land	92,650	0.12	1.0	2	South Africa	Pilot	Operational	0.930
Lotte CCUS Project	Offshore (−1.11, 54.58)	/	/	/	/	/	/	/	UK	Pilot	In Design	0
Otway Basin	Onshore (150.63, −24.33)	Low coverage	Low good slope	Impervious surface	81,750	0.075	2.0	10	Australia	Pilot	Finished	0.842
PEMEX EOR Pilot	Onshore (−94.56, 18.00)	Medium coverage	Medium slope	Impervious surface	50,000	0.5	1.8	1	Mexico	Commercial	In Design	0.522
Ras Laffan	Onshore (51.54, 25.91)	Barren	Low good slope	Bare land	300,000	2.1	1.0	1	Qatar	Commercial	Operational	0.656
RECOPOL	Onshore (16.84, 51.47)	Medium-low	Medium slope	Cropland	510	0.0001	1.1	1	Poland	Pilot	Operational	0.699
ROAD	Offshore (4.02, 51.96)	/	/	/	/	/	/	/	Netherlands	Pilot	Finished	0
San Juan	Onshore (−108.44, 36.80)	Barren	Very low good slope	Bare land	2100	0.104	0.8	5	USA	Commercial	Finished	0.815
Shenhua Ordos Pilot	Onshore (110.15, 39.33)	Low coverage	Low good slope	Impervious surface	60,000	0.3	2.7	3	China	Pilot	Finished	0.632
Sleipner	Offshore (3.00, 58.41)	/	/	/	/	/	/	/	Norway	Commercial	Finished	0
Tomakomai CCS Project	Offshore (141.65, 42.63)	/	/	/	/	/	/	/	Japan	Commercial	Finished	0

Table 10 *Cont.*

Project Name	Reservoir Location	Vegetation Coverage	Topographic Factor	LULC	Injection Rate (m^3/Day)	Total Injection (Mt)	Reservoir Depth (km)	Duration (Year)	Country	Scale	Status	F_{Insar}
Lacq	Onshore (−0.67, 43.44)	Low coverage	Very low good slope	Grassland	109,000	0.012	4.5	2	France	Commercial	Operational	0.843
Uthmaniyah EOR Project	Onshore (49.36, 24.80)	Barren	Very low good slope	Eare land	1,274,000	4.0	3.6	5	Saudi Arabia	Pilot	Operational	0.883
West Texas Scurry field	Onshore (−101.09, 32.07)	Low coverage	Low good slope	Grassland	90,000	55.0	2.0	35	USA	Commercial	Finished	0.712
Weyburn CO_2 Project	Onshore (−103.68, 49.51)	Barren	Low good slope	Grassland	7,085,000	25.0	1.5	15	Canada	Commercial	Operational	0.832
Yangchang Jingbian	Onshore (108.91, 37.42)	Low coverage	Medium slope	Grassland	7000	0.005	3.0	5	China	Pilot	Finished	0.694
Youngil Bay	Offshore (129.46, 36.06)	/	/	/	/	/	/	/	South Korea	Pilot	Finished	0

5. Discussion

A CCS project is one of the most effective methods to reduce carbon emissions. Whether it can effectively monitor storage leakage has become one of the determinants of its large-scale commercialization. Part of the CO_2 storage in the geological layer is fixed in the storage layer through physical and chemical action, another part flows along the geological layer, and a small part will penetrate or leak to the soil, atmosphere, and other places along the geological defects, wells, and other parts.

CO_2 storage monitoring can be divided into two types: leakage monitoring during injection and long-term migration monitoring after injection. The purpose of leakage monitoring is to prevent the ecological environment damage caused by the overflow of CO_2 during the injection process. The long-term migration monitoring after injection is to explore the migration pathway of CO_2 and the front of CO_2 migration. In the monitoring of the injection process, the direct measurement equipment on the surface and underground is a better choice. For long-term monitoring, a spaceborne InSAR sensor is obviously a better choice due to the limitation of cost and labor.

However, InSAR technology still faces many challenges for CCS monitoring. As an indirect monitoring technology, InSAR monitoring can only detect the migration pathway of CO_2 through the response of the surface. Geological data or other monitoring data will be a good supplement and verification. Therefore, on the premise of InSAR monitoring, combined with geological, logging, and other methods of joint monitoring will be the development trend of monitoring in the future. Figure 5 is a feasible multi-method monitoring system. Through underground stress analysis and a well monitoring subsystem, the distribution and migration of CO_2 can be determined intuitively, but the result is point distribution. The underground fractures trend and energy distribution can be analyzed by microseismic and other surface monitoring subsystems, so as to obtain planar migration results. The aerospace monitoring subsystem of InSAR and UAV can monitor CO_2 migration for a long time and obtain the overall dynamic process of CO_2 migration. Various monitoring subsystems complement and confirm each other, efficiently and accurately detect the migration pathway, and realize all-round migration and leakage safety monitoring.

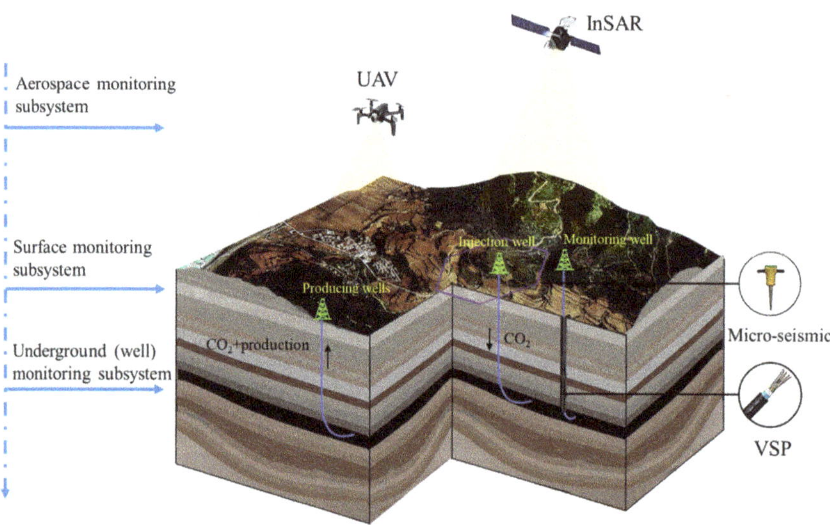

Figure 5. A multi-method monitoring system.

In this study, we selected eight factors affecting InSAR monitoring, mainly considering whether InSAR can effectively monitor the surface deformation caused by CO_2 injection. The causes of surface deformation at the storage site are usually diverse. The common interference factors include deformation caused by tectonic activity, production processes, and human activities. The evaluation method we proposed includes most cases as much as possible. For the tectonic activity factors, detailed risk assessment will be conducted during the storage and site selection process. Therefore, the possibility of large-scale tectonic movement in this area is very small, and the stable small-scale surface deformation can be removed through the background value. Therefore, the deformation composed of tectonic activity will not affect the InSAR's monitoring of surface deformation caused by CO_2 injection. For different storage types, the response of surface deformation will be different due to the different lithology of reservoir and caprock, but the response of large-scale commercial storage volume is enough to be detected by InSAR. Therefore, this study only divides the storage types into onshore and offshore. For the corresponding relationship between deformation and different reservoirs, it is necessary to analyze the specific storage site. The deformation caused by the production process is characterized by short deformation time and large deformation value. It can be separated by filtering the time series results, so it will not interfere with the InSAR results. The biggest interference of human activities on InSAR deformation detection is the planting of vegetation or crops. This situation has been discussed in LULC, so it will not be mentioned again.

In addition, the possibilities of leakage are also one of the important factors determining the necessity of monitoring. Since the site selection process of CO_2 storage has been carefully evaluated, geological factors are often not the main factor determining the injection leakage risk. The injection method, injection rate, and injection volume also determine the risk of leakage to a certain extent. These factors are reflected in our evaluation model, but some factors are not considered. For example, too many wells also mean more leakage risk, and the sudden change of underground stress will also lead to leakage. However, these factors are not decisive factors and are not included in this evaluation model. Therefore, it is possible to further improve the details of the model in the future.

6. Conclusions

The feasibility assessment of InSAR is a necessary and important step for CCS monitoring. We proposed a new feasibility evaluation model based on the properties of InSAR technology and the characteristics of the sequestration areas. Eight factors were selected as parameters to conduct the feasibility assessment using InSAR in an equal weight method. In addition, this study evaluated the typical completed or ongoing sequestration projects around the world and utilized the research areas that had successful sequestration for verification. Suggestions on whether InSAR technology could be used for monitoring were also provided. Based on the results presented, the following conclusions were reached:

(1) InSAR technology is a potential monitoring method for CCS sequestration leakage and migration, but it is still in its infancy. More than 130 CCS sequestration areas were utilized in this study, but less than 10 were monitored using the InSAR technique. In addition, the results were also uneven. The reasons are complex and varied, but they can be divided into the characteristic limitations of InSAR technology and the special requirements of CCS sequestration. Therefore, a parameter is required to assess whether InSAR monitoring is viable for a specific region.

(2) InSAR technology still has technical limitations for monitoring surface deformation, especially in complex mountainous areas. Due to safety considerations, CCS storage areas are often built far away from cities, including mountains and deserts. These places are inevitably faced with layout, foreshortening, and other incoherence phenomena, resulting in the absence and inaccuracy of results. In this study, the fractional vegetation cover and R-index were selected to evaluate the influence of vegetation and topography on the results of InSAR, so as to evaluate the feasibility of InSAR.

(3) The use of InSAR technology also has certain requirements for the injection mode and the injection area. According to limitations in the field design and the injection plan, the injection rate and injection amount of each storage area were different. However, a slow and small amount of injection had an extremely insignificant and irregular response on the surface deformation, and the regular deformation trend could not be identified. Hence, it could not be determined whether CO_2 injection was the primary controlling factor of the surface deformation. Therefore, we evaluated the injection volume, injection depth, injection rate, land use/land cover, reservoir type, and monitoring duration. Given the fixed evaluation criteria, we concluded that a large amount and stable injection were suitable for deformation monitoring, so as to realize the feasibility evaluation of InSAR technology in CO_2 sequestration areas.

(4) To comprehensively evaluate the feasibility of InSAR monitoring in CO_2 sequestration areas, an evaluation model of the InSAR feasibility was established according to the characteristics of InSAR and the CCS sequestration areas. For a typical injection area or under injection areas, suggestions were made on whether InSAR can be used for monitoring. In addition, the cases of successful application and failure to detect deformation were verified. The cases of successful monitoring and the cases wherein deformation was not accurately detected were both used to validate the evaluation parameters. The results demonstrated that most of the sequestration areas were suitable for the InSAR technique, and the feasibility was consistent with the collected data.

It must be noted that the feasibility assessment equation was based on the successful monitoring cases of InSAR that are currently available. Since there were few cases available for analysis, the weight of each factor requires further consideration. With improvement in the InSAR method, the evaluation method can be changed. Currently, this assessment method is generally applicable to time-series InSAR technologies, such as persistent scatterer InSAR (PS-InSAR) and small baseline subset InSAR (SBAS-InSAR). Similarly, with the improvement in SAR accuracy, different bands of radar will also affect the assessment. The corresponding evaluation model needs to be constantly updated with the technology. However, InSAR technology is worthy of popularization and has extensive application prospects in the field of CCS monitoring in the future.

Author Contributions: T.Z. designed this study. T.Z. performed the data collection, processing, and analysis. D.C. and H.G. optimized the figure for this work. R.Y. provided CCS data for this work. The corresponding author W.Z. is supervisor of this work and contributed with continuous guidance during this work. T.Z. jointly wrote this manuscript, and the manuscript was edited by W.Z. All authors have read and agreed to the published version of the manuscript.

Funding: This study was jointly financed by the National Key R&D Program of China (Grant No. 2018YFB0605603).

Acknowledgments: The authors are grateful to the anonymous reviewers for their constructive comments and suggestions to improve this manuscript.

Conflicts of Interest: The authors declare no conflict of interest.

References

1. Friedlingstein, P.; O'Sullivan, M.; Jones, M.W.; Andrew, R.M.; Hauck, J.; Olsen, A.; Peters, G.P.; Peters, W.; Pongratz, J.; Sitch, S.; et al. Global Carbon Budget 2020. *Earth Syst. Sci. Data* **2020**, *12*, 3269–3340. [CrossRef]
2. Tapia, J.F.D.; Lee, J.Y.; Ooi, R.E.H.; Foo, D.C.Y.; Tan, R.R. A review of optimization and decision-making models for the planning of CO_2 capture, utilization and storage (CCUS) systems. *Sustain. Prod. Consum.* **2018**, *13*, 1–15. [CrossRef]
3. Zendehboudi, S.; Khan, A.; Carlisle, S.; Leonenko, Y. Ex situ dissolution of CO_2: A new engineering methodology based on mass-transfer perspective for enhancement of CO_2 sequestration. *Energy Fuels* **2011**, *25*, 3323–3333. [CrossRef]
4. Du, J.; Brissenden, S.J.; Mcglllivray, P.; Bourne, S.J.; Hofstra, P.; Davis, E.J.; Roadarmel, W.M.; Wolhart, S.L.; Wright, C.A. Mapping fluid flow in a reservoir using tiltmeter-based surface-deformation measurements. In Proceedings of the SPE Annual Technical Conference and Exhibition, Dallas, TX, USA, 9 October 2005; pp. 25–32. [CrossRef]

5. IEAGHG. *Monitoring and Modelling of CO$_2$ Storage: The Potential for Improving the Cost-Benefit Ratio of Reducing Risk*; IEAGHG Technical Report 2020–01; IEAGHG: Cheltenham, UK, February 2020.

6. Zhang, T.; Zhang, W.; Yang, R.; Liu, Y.; Jafari, M. CO$_2$ capture and storage monitoring based on remote sensing techniques: A review. *J. Clean. Prod.* **2021**, *281*, 124409. [CrossRef]

7. Raziperchikolaee, S.; Cotter, Z.; Gupta, N. Assessing mechanical response of CO$_2$ storage into a depleted carbonate reef using a site-scale geomechanical model calibrated with field tests and InSAR monitoring data. *J. Nat. Gas Sci. Eng.* **2021**, *86*, 103744. [CrossRef]

8. Loschetter, A.; Rohmer, J.; Raucoules, D.; De Michele, M. Sizing a geodetic network for risk-oriented monitoring of surface deformations induced by CO$_2$ injection: Experience feedback with InSAR data collected at In-Salah, Algeria. *Int. J. Greenh. Gas Control* **2015**, *42*, 571–582. [CrossRef]

9. Peltier, A.; Froger, J.L.; Villeneuve, N.; Catry, T. Assessing the reliability and consistency of InSAR and GNSS data for retrieving 3D-displacement rapid changes, the example of the 2015 Piton de la Fournaise eruptions. *J. Volcanol. Geotherm. Res.* **2017**, *344*, 106–120. [CrossRef]

10. Radutu, A.; Nedelcu, I.; Gogu, C.R. An overview of ground surface displacements generated by groundwater dynamics, revealed by InSAR techniques. *Procedia Eng.* **2017**, *209*, 119–126. [CrossRef]

11. Bohloli, B.; Bjørnarå, T.I.; Park, J.; Rucci, A. Can we use surface uplift data for reservoir performance monitoring? A case study from In Salah, Algeria. *Int. J. Greenh. Gas Control* **2018**, *76*, 200–207. [CrossRef]

12. Rutqvist, J.; Vasco, D.W.; Myer, L. Coupled reservoir-geomechanical analysis of CO$_2$ injection and ground deformations at In Salah, Algeria. *Int. J. Greenh. Gas Control* **2010**, *4*, 225–230. [CrossRef]

13. Mathieson, A.; Wright, I.; Roberts, D.; Ringrose, P. Satellite imaging to monitor CO$_2$ movement at Krechba, Algeria. *Energy Procedia* **2009**, *1*, 2201–2209. [CrossRef]

14. Vasco, D.W.; Rucci, A.; Ferretti, A.; Novali, F.; Bissell, R.C.; Ringrose, P.S.; Mathieson, A.S.; Wright, I.W. Satellite-based measurements of surface deformation reveal fluid flow associated with the geological storage of carbon dioxide. *Geophys. Res. Lett.* **2010**, *37*. [CrossRef]

15. Yunjun, Z.; Fattahi, H.; Amelung, F. Small baseline InSAR time series analysis: Unwrapping error correction and noise reduction. *Comput. Geosci.* **2019**, *133*, 104331. [CrossRef]

16. Onuma, T.; Ohkawa, S. Detection of surface deformation related with CO$_2$ injection by DInSAR at In Salah, Algeria. *Energy Procedia* **2009**, *1*, 2177–2184. [CrossRef]

17. Yang, Q.; Zhao, W.; Dixon, T.H.; Amelung, F.; Han, W.S.; Li, P. InSAR monitoring of ground deformation due to CO$_2$ injection at an enhanced oil recovery site, West Texas. *Int. J. Greenh. Gas Control* **2015**, *41*, 20–28. [CrossRef]

18. Guo, S.; Zheng, H.; Yang, Y.; Zhang, S.; Hou, H.; Zhu, Q.; Du, P. Spatial estimates of surface deformation and topsoil moisture in operating CO$_2$-EOR project: Pilot environmental monitoring using SAR technique. *J. Clean. Prod.* **2019**, *236*, 117606. [CrossRef]

19. Samsonov, S.; Czarnogorska, M.; White, D. Satellite interferometry for high-precision detection of ground deformation at a carbon dioxide storage site. *Int. J. Greenh. Gas Control* **2015**, *42*, 188–199. [CrossRef]

20. Shi, J.; Yang, H.; Peng, J.; Wu, L.; Xu, B.; Liu, Y.; Zhao, B. InSAR Monitoring and Analysis of Ground Deformation Due to Fluid or Gas Injection in Fengcheng Oil Field, Xinjiang, China. *J. Indian Soc. Remote Sens.* **2019**, *47*, 455–466. [CrossRef]

21. Li, Z.; Duan, M.; Cao, Y.; Mu, M.; He, X.; Wei, J. Mitigation of time-series InSAR turbulent atmospheric phase noise: A review. *Geod. Geodyn.* **2022**, 1–11. [CrossRef]

22. Gabriel, A.K.; Goldstein, R.M.; Zebker, H.A. Mapping small elevation changes over large areas: Differential radar interferometry. *J. Geophys. Res.* **1989**, *94*, 9183–9191. [CrossRef]

23. Bai, Z.; Fang, S.; Gao, J.; Zhang, Y.; Jin, G.; Wang, S.; Zhu, Y.; Xu, J. Could Vegetation Index be Derive from Synthetic Aperture Radar?—The Linear Relationship between Interferometric Coherence and NDVI. *Sci. Rep.* **2020**, *10*, 6749. [CrossRef]

24. Gao, L.; Wang, X.; Johnson, B.A.; Tian, Q.; Wang, Y.; Verrelst, J.; Mu, X.; Gu, X. Remote sensing algorithms for estimation of fractional vegetation cover using pure vegetation index values: A review. *ISPRS J. Photogramm. Remote Sens.* **2020**, *159*, 364–377. [CrossRef]

25. Jia, K.; Liang, S.; Liu, S.; Li, Y.; Xiao, Z.; Yao, Y. Global Land Surface Fractional Vegetation Cover Estimation Using General Regression Neural Networks From MODIS Surface Reflectance. *IEEE Trans. Geosci. Remote Sens.* **2015**, *53*, 4787–4796. [CrossRef]

26. Colesanti, C.; Wasowski, J. Investigating landslides with space-borne Synthetic Aperture Radar (SAR) interferometry. *Eng. Geol.* **2006**, *88*, 173–199. [CrossRef]

27. Cigna, F.; Bateson, L.B.; Jordan, C.J.; Dashwood, C. Simulating SAR geometric distortions and predicting Persistent Scatterer densities for ERS-1/2 and ENVISAT C-band SAR and InSAR applications: Nationwide feasibility assessment to monitor the landmass of Great Britain with SAR imagery. *Remote Sens. Environ.* **2014**, *152*, 441–466. [CrossRef]

28. Notti, D.; Davalillo, J.C.; Herrera, G.; Mora, O. Assessment of the performance of X-band satellite radar data for landslide mapping and monitoring: Upper Tena Valley case study. *Nat. Hazards Earth Syst. Sci.* **2010**, *10*, 1865–1875. [CrossRef]

29. Dai, Z.; Viswanathan, H.; Xiao, T.; Middleton, R.; Pan, F.; Ampomah, W.; Yang, C.; Zhou, Y.; Jia, W.; Lee, S.Y.; et al. CO$_2$ Sequestration and Enhanced Oil Recovery at Depleted Oil/Gas Reservoirs. *Energy Procedia* **2017**, *114*, 6957–6967. [CrossRef]

30. Mukherjee, M.; Misra, S. A review of experimental research on Enhanced Coal Bed Methane (ECBM) recovery via CO$_2$ sequestration. *Earth-Sci. Rev.* **2018**, *179*, 392–410. [CrossRef]

31. Jayasekara, D.W.; Ranjith, P.G.; Wanniarachchi, W.A.M.; Rathnaweera, T.D. Understanding the chemico-mineralogical changes of caprock sealing in deep saline CO_2 sequestration environments: A review study. *J. Supercrit. Fluids* **2020**, *161*, 104819. [CrossRef]
32. Li, Q.; Wu, Z.; Li, X. Prediction of CO_2 leakage during sequestration into marine sedimentary strata. *Energy Convers. Manag.* **2009**, *50*, 503–509. [CrossRef]
33. Xiong, Z.; Li, T.; Crosta, X.; Algeo, T.; Chang, F.; Zhai, B. Potential role of giant marine diatoms in sequestration of atmospheric CO_2 during the Last Glacial Maximum: δ13C evidence from laminated Ethmodiscus rex mats in tropical West Pacific. *Glob. Planet. Change* **2013**, *108*, 1–14. [CrossRef]
34. Ren, Y.; Lü, Y.; Comber, A.; Fu, B.; Harris, P.; Wu, L. Spatially explicit simulation of land use/land cover changes: Current coverage and future prospects. *Earth-Sci. Rev.* **2019**, *190*, 398–415. [CrossRef]
35. NRSCC. *Global Ecosystems and Environment Observation Analysis Report Cooperation (GEOARC) 2017*; NRSCC: Beijing, China, 2017.
36. Cao, D.; Zhang, J.; Xun, L.; Yang, S.; Wang, J.; Yao, F. Spatiotemporal variations of global terrestrial vegetation climate potential productivity under climate change. *Sci. Total Environ.* **2021**, *770*, 145320. [CrossRef]
37. Sreejith, K.M.; Agrawal, R.; Agram, P.; Rajawat, A.S. Surface deformation of the Barren Island volcano, Andaman Sea (2007–2017) constrained by InSAR measurements: Evidence for shallow magma reservoir and lava field subsidence. *J. Volcanol. Geotherm. Res.* **2020**, *407*, 107107. [CrossRef]
38. Abdel-Hamid, A.; Dubovyk, O.; Greve, K. The potential of sentinel-1 InSAR coherence for grasslands monitoring in Eastern Cape, South Africa. *Int. J. Appl. Earth Obs. Geoinf.* **2021**, *98*, 102306. [CrossRef]
39. Teng, Y.; Liu, Y.; Lu, G.; Jiang, L.; Wang, D.; Song, Y. Experimental Evaluation of Injection Pressure and Flow Rate Effects on Geological CO_2 Sequestration Using MRI. *Energy Procedia* **2017**, *114*, 4986–4993. [CrossRef]
40. Zheng, Y.; Liu, J.; Zhang, B. The Influence of Geological Storage of Carbon Dioxide on Deformation of Ground Surface in Depleted Reservoirs. *J. Hebei Geo Univ.* **2019**, *2*, 1–6. [CrossRef]
41. Schaef, H.T.; McGrail, B.P.; Owen, A.T. Basalt reactivity variability with reservoir depth in supercritical CO_2 and aqueous phases. *Energy Procedia* **2011**, *4*, 4977–4984. [CrossRef]
42. Murdoch, L.C.; Germanovich, L.N.; DeWolf, S.J.; Moysey, S.M.J.; Hanna, A.C.; Kim, S.; Duncan, R.G. Feasibility of using in situ deformation to monitor CO_2 storage. *Int. J. Greenh. Gas Control* **2020**, *93*, 102853. [CrossRef]
43. Hannis, S.; Lu, J.; Chadwick, A.; Hovorka, S.; Kirk, K.; Romanak, K.; Pearce, J. CO_2 Storage in Depleted or Depleting Oil and Gas Fields: What can We Learn from Existing Projects? *Energy Procedia* **2017**, *114*, 5680–5690. [CrossRef]

Article

Preparation, Properties and Application of Gel Materials for Coal Gangue Control

Xiaoqiang Zhang * and Yuanyuan Pan

College of Environment and Resources, Xiangtan University, Xiangtan 411100, China; panyuanyuan202109@163.com
* Correspondence: xiaoqiangzhang@xtu.edu.cn

Abstract: In order to solve the problem of the spontaneous combustion of coal gangue, a coal gangue fire-extinguishing material of gel–foam was developed. The foaming agent was screened by the Waring blender method with varying foam amounts, and the superabsorbent foam stabilizer was synthesized by free radical polymerization. Moreover, the gel–foam was used in a spontaneous combustion of coal gangue mountain field practice. The results showed that when the mass fraction of sodium dodecyl sulfonate and coconut oil amide propyl betaine was 0.6% and 4:6, the foaming amount was as high as 1500 mL. When the mass ratio of chitosan to acrylic acid was 1:6, the neutralization degree was 80%, the cross-linking agent was 0.8%, and the initiator was 0.01%, the water absorption of the synthesized superabsorbent foam stabilizer reached 476 mL/g. The synthesized gel–foam was tested in a spontaneous combustion coal gangue hill in a certain area, and no reburning sign was found within one month.

Keywords: gel–foam; chitosan; acrylic acid; coal gangue

Citation: Zhang, X.; Pan, Y. Preparation, Properties and Application of Gel Materials for Coal Gangue Control. *Energies* **2022**, *15*, 557. https://doi.org/10.3390/en15020557

Academic Editor: Nikolaos Koukouzas

Received: 6 December 2021
Accepted: 6 January 2022
Published: 13 January 2022

Publisher's Note: MDPI stays neutral with regard to jurisdictional claims in published maps and institutional affiliations.

1. Introduction

Coal is the most commonly used energy source at present; consequently, coal mining brings a large amount of coal gangue emissions. The most commonly used treatment method for coal gangue is natural accumulation, to form a coal gangue mountain. China is a big country in coal production and consumption; according to incomplete statistics, there are nearly 2000 large coal gangue hills, with about 6 billion tons of accumulated coal gangue, covering an area of about 20,000 hectares, and increasing by 150–200 million tons per year [1–4]. The overall distribution of coal gangue is characterized by "more in the north than in the south, and less in the west than in the east". The national coal gangue emission distributions are shown in Figure 1. Combined with the occurrence status and regional distribution characteristics of coal resources in China, considering the coal gangue horizon and the coal accumulation basin, coal gangue production areas in China can be roughly divided into five regions: namely, Northeast China, North China, South China, the Southwest, and the Northwest. According to the output of the producing area, they are divided into five grades of red, orange, yellow, green, and blue, as shown in Figure 2 below [5,6].

Coal gangue contains about 15% of coal and other fuels. Due to the wind blowing and raining all year-round, it is naturally oxidized in a certain environment and generates heat [7–9]. Under sufficient oxygen and a good heat storage environment, spontaneous combustion is likely to occur when it reaches an ignition point [10]. At present, more than one third of the 2000 large coal gangue mines in China have experienced, or are even currently experiencing, spontaneous combustion. This produces a large number of toxic and harmful gases such as CO, CO_2, SO_2 and NO_x, which are discharged into the atmospheric environment, and heavy metal elements such as arsenic, mercury, fluorine, lead and selenium are released into the soil and groundwater [11–13]. This pollutes the surrounding atmosphere, water and soil, causes serious environmental and ecological pollution problems, and seriously threatens the production safety of mining areas and the health of surrounding personnel [14]. A fire prevention and extinguishing material was prepared for the spontaneous combustion of coal gangue at present.

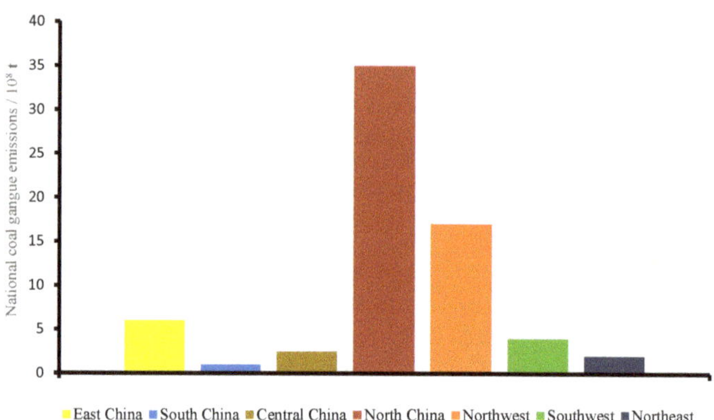

Figure 1. Cumulative emissions of coal gangue from various regions of the country.

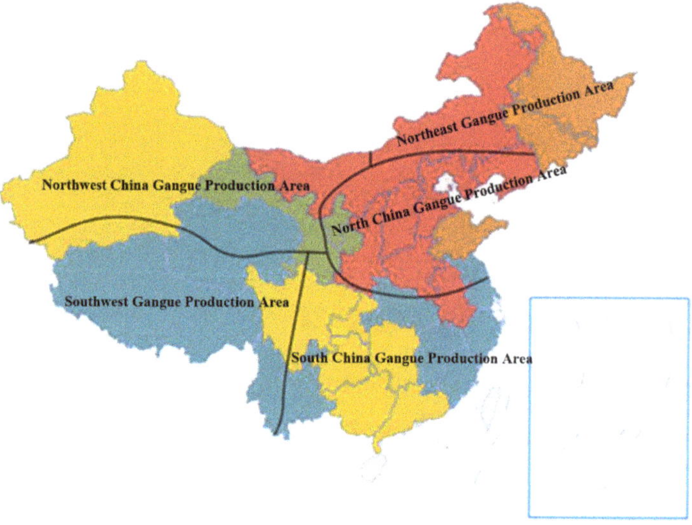

Figure 2. Distribution of gangue producing areas in China.

Domestic scholars have drawn lessons from many fire-extinguishing materials in coal gangue fire prevention and extinguishing, and water-based foam is the most widely used fire-extinguishing material. However, the interface of water-based foam can only be stabilized with the help of surfactants, which are easy to collapse when transported through porous media (such as clots). Therefore, water-based foam is not suitable for long-term fire prevention [15,16]. Compared with water-based foam, three-phase foam has better gas resistance, greater residual resistance and a longer attenuation half-life. However, the stability of three-phase foam is limited by rising temperatures, resulting in fire failure. At present, gel has been used as an external phase to improve the stability of the foam, called gel–foam. After gelation, bubbles are firmly trapped in the high viscosity gel film in porous media, improving its resilience, heat absorption and blocking ability, thereby reducing the flow of oxygen [17]. Gel–foam, using foam as a carrier, can be transported to higher and farther places, as foam itself has a certain sealing performance. A large flow of gel–foam can quickly accumulate, and ignition gel has good cooling and oxygen insulation. In addition, when the foam bursts, the foamed gel can still maintain a porous structure and respond quickly to temperature changes, resulting in rapid water loss at high temperatures. Finally, the gel can encapsulate and seal the high-temperature fire source. High water content can keep coal in a humid environment, and through its cooling, wetting and ability to prevent floating coal from becoming exposed to oxygen, the spontaneous combustion of coal can be avoided. Therefore, the combination of gel fire-extinguishing technology and foam fire-extinguishing technology not only has the advantages of gel fire-extinguishing, but also the low cost of foam fire-extinguishing. The research on gel–foam has attracted more and more attention, and the development of a new gel–foam fire-extinguishing material for preventing the spontaneous combustion of coal gangue has broad application prospects [18].

Gel–foam is a kind of complex mixed system, which is formed by dispersing the polymer in water, adding a foaming agent and foaming under the action of gas. After a period of time, in the foam–liquid film, the polymers crosslink each other to form a three-dimensional network structure, which constitutes the rigid skeleton of gel–foam. Gel–foam has special properties, including not only the properties of gel, but also the characteristics of foam. It has good temperature resistance, fire resistance, plugging and inhibition performance, good film-forming ability, long-term effective coating of coal, and prominent reburning prevention effect. The foam is used as the carrier of gel, and the perfusion of fire-extinguishing materials has been improved to a certain extent. The fire-extinguishing efficiency is high, and the control effect on coal spontaneous combustion in small-scale high-rise areas is obvious [19–21].

2. Experimental Part

2.1. Main Experimental Materials

(1) Foaming agent material. Eight representative and economically appropriate surfactants were selected and numbered as 1 #–8 #, respectively. The materials are shown in Table 1;

Table 1. Types of surfactants.

Classification of Surfactants	Surfactant Types
Cationic surfactant	Cetyltrimethylammonium bromide (CTAB)
Anionic surfactant	Sodium dodecyl sulfonate (SDS) Sodium dodecyl benzene sulfonate (SDBS) Sodium dodecyl sulfonate (SAS)
Amphoteric surfactant	Cocoamidopropyl betaine (CB) Sodium polyoxyethylene dodecyl ether sulfate (AES) Sulfonate betaine (SB) Imidazoline phosphate amphoteric surfactant (FP2)

(2) Superabsorbent foam-stabilizer monomer: chitosan, acrylic acid;
(3) Crosslinkers and initiators:

Cross-linking agent: N, N′-methylene bisacrylamide;
Initiator: potassium persulfate.

2.2. Synthesis of Superabsorbent Foam Stabilizer

In this experiment, superabsorbent foam stabilizer was synthesized by a hydrothermal method. A certain mass of chitosan was dissolved in 2% glacial acetic acid solution and stirred evenly in a magnetic stirrer, until it became a viscous liquid. Then, a certain amount of potassium persulfate solution was added, and the neutralization degree of acrylic acid reached 80% by neutralizing acrylic acid solution, with 20% sodium hydroxide solution in an ice bath. After the temperature of acrylic acid solution was restored to room temperature, the above solution was transferred into chitosan solution and stirred evenly. Then, a certain mass of cross-linking agent (N, N′-methylene bisacrylamide) was added, and the gel was prepared by heating the magnetic stirrer in a water bath with a heating constant temperature. The reaction temperature was controlled, and the gel was protected by nitrogen. The main test process is shown in Figure 3a.

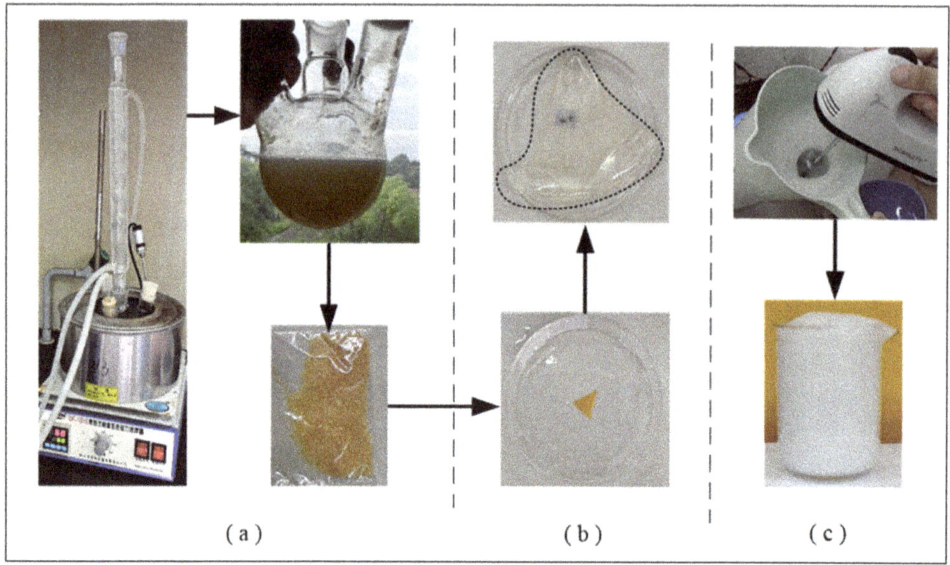

Figure 3. Schematic diagram of experimental process. (**a**) preparation of superabsorbent foam stabilizer. (**b**) water absorption test of foam stabilizer sample. (**c**) preparation of gel foam.

2.3. Experimental Method

(1) Foaming experiment. The Waring blender stirring–foaming method was used. The rotation speed of the agitator reached 1000 r/min. When the stirring was longer than 60 s, the foam volume number was immediately stopped and read. The numerical value was the foaming capacity of the material. Figure 3c shows the surfactant stirring foaming process;

(2) Water absorption performance test experiment. The natural filtration method was used to determine the water absorption rate. Firstly, 1 g superabsorbent gel sample was accurately weighed, placed in a 1000 mL beaker, and a certain volume (1000 mL) of distilled water was added (this material was used in a coal-mine production practice, so the water-absorption rate test was still carried out with tap water). After its water absorption was saturated, the free water was filtered with 200-mesh nylon cloth, and placed on it for 30 min. Then the sample quality was weighed. It is shown the swelling process of super absorbent foam stabilizer samples in Figure 3b;

(3) Stability experiment. There are two main methods to evaluate foam stability. One is the defoaming half-life, which means the time needed to reduce the foam volume by half, and is called the bursting half-life. The other is the liquid-separation half-life, which means the time needed to separate surfactant solution by half, and is called the liquid separation half-life. In this paper, the defoaming half-life method was used to evaluate the foam stability of the foam stabilizer;

(4) Temperature resistance experiment. Using an electric blast drying oven as a heating source, we set the heating temperature to 200 °C unchanged. The temperature resistance time of different gel–foam ratios was investigated according to the half-life.

3. Experiment Process and Result Analysis

Gel–foam is a dispersion system composed of the average distribution of gas in the gel. It is generally composed of three parts: one is a foaming agent that can produce foam; one is a foam stabilizer that can prolong the foam persistence; the other is a gel forming system that constitutes the foam skeleton structure. In this part, surfactant was selected by a single compounding method, and then the superabsorbent foam stabilizer was synthesized, and the temperature resistance of the gel–foam was tested.

3.1. Selection of Foaming Agent

(1)   Selection of single foaming agent

Based on the principle of the comprehensive performance of the foaming agents, eight surfactants were selected for foaming capacity experiments. Firstly, the foaming capacity of a single foaming agent was measured by the Waring blender method, and three materials with the best starting foaming ability were selected for eight materials by the Waring blender stirring method. The standard waring-blender method uses distilled water. Since this product will be used in coal mine production, tap water was used in this experiment. The foaming volume of each foaming agent is shown in Figure 4.

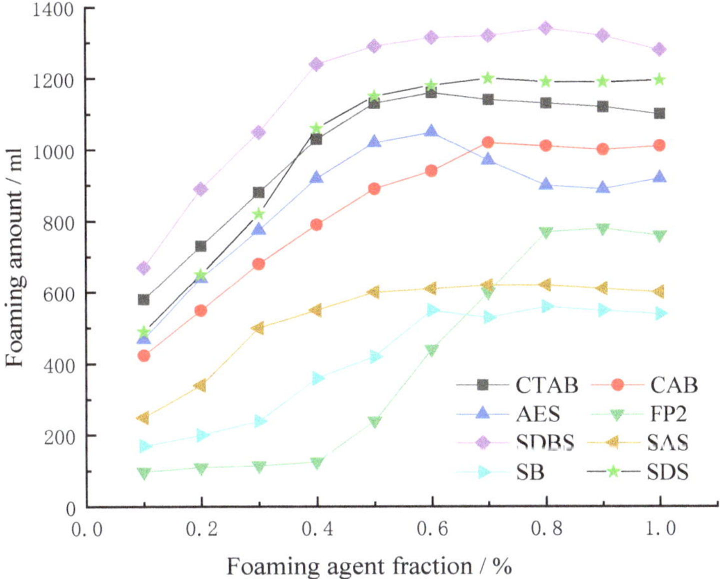

Figure 4. Foaming multiples of a single foaming agent.

In the Waring blender method, the foam is produced by the liquid under mechanical agitation. However, the factors affecting the foaming capacity of the liquid are more complicated. The concentration of the foaming agent is the most important influencing factor. As can be seen from Figure 4, the foaming volume increases with the increase of surfactant concentration in a certain concentration range, but when the foaming agent concentration reaches a certain value, the trend of the foaming volume tends to level off. This is because, at the turning point, the concentration of foaming agent has reached its critical micelle concentration. If the concentration of foaming agent continues to increase, the foaming agent can only form micelles in the liquid, and cannot continue to reduce the surface tension of the foaming system.

At the same concentration, SDS, SDBS, CAB, AES have stronger foaming abilities, and the maximum foaming volume is more than 1000 mL. This is mainly because the effective surface tension of SDS, SDBS, CAB and AES is lower than that of SAS, SB and FP2. Effective surface tension is characterized by a lower surface tension that is conducive to reducing the surface energy of the liquid, and conducive to increasing the foam volume of the blowing agent. In addition, because water has a certain degree of hardness, the foaming volume of the foaming agent with poor hard-water resistance is small. Four kinds of foaming agents, SDS, SDBS, CAB and AES, which have better foaming properties, were selected for pairwise compounding experiments.

(2)   Compounding of foaming.

We combined SDBS, CAB, AES, CAB, four kinds of foaming agents, into pairs; the concentration was determined to be 0.6%. Then, we changed the ratio of the four kinds of foaming agents when paired, and used the Waring blender method to carry out the foaming experiment, and obtained the foam. The volume change is shown in Figure 5.

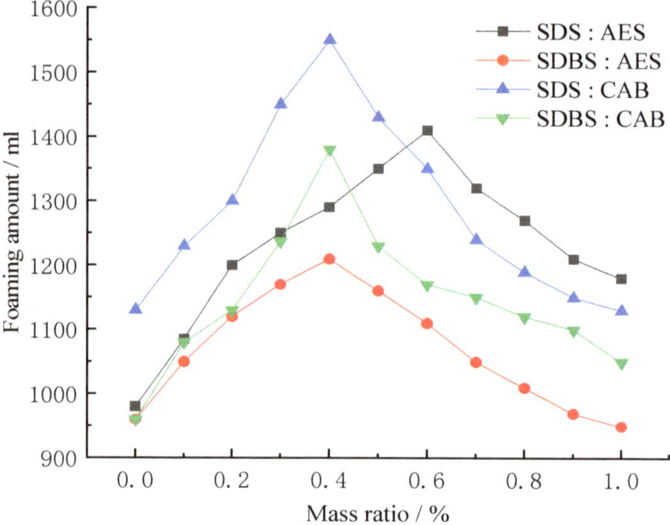

Figure 5. Foaming multiple of pairwise compounding foaming agent.

As can be seen from Figure 5, the foaming volume of the composite foaming agent increased with the approach of the relative proportion of foaming agents, and was significantly higher than that of the single foaming agent. Overall, a synergistic effect was achieved by using two blowing agents in appropriate proportions. Considering the economic and effective factors, 0.6% volume fraction with SDS:CAB ratio of 4: 6 was chosen as the best.

3.2. Preparation of Foam Stabilizer

(1)   Optimization of thermal preparation conditions of chitosan-acrylic acid highly absorbent hydrogels.

The ratio of chitosan and acrylic acid monomer, initiator concentration, cross-linking agent concentration and neutralization degree of acrylic acid in chitosan-grafting acrylic acid have great influence on the water absorption of foam stabilizer. Through orthogonal experiment, the chitosan acrylic acid mass fraction, chitosan concentration, cross-linking agent concentration and neutralization degree of acrylic acid were optimized. The orthogonal factor levels are shown in Table 2, and orthogonal analysis data are shown in Table 3. A is the ratio of volume of acrylic acid to mass of chitosan (mL/g), B is initiator mass concentration (%), C is mass concentration of the cross-linking agent (%), and D is neutralization degree of acrylic acid (%).

Table 2. Orthogonal level factor table.

Level	A	B	C	D
1	2	0.4	0.01	40
2	4	0.6	0.03	60
3	6	0.8	0.05	80

Table 3. Analysis of orthogonal experimental results for synthesis of superabsorbent hydrogel.

Experiment Number	Experimental Factors				Water Absorbency (mL/g)
	A	B	C	D	
1	2	0.4	0.03	80	205
2	2	0.8	0.05	40	53
3	2	1.2	0.01	60	122
4	4	0.4	0.01	40	86
5	4	0.8	0.03	60	152
6	4	1.2	0.05	80	288
7	6	0.4	0.05	60	235
8	6	0.8	0.01	80	476
9	6	1.2	0.03	40	120
K1	126.67	175.33	228	86.34	
K2	175.33	227	159	169.67	
K3	277	176.67	192	323	
Range	150.33	51.67	69	236.67	
Important order	D > A > C > B				
Optimization levels	A_3	B_2	C_1	D_3	
Superior combination	$A_3B_2C_1D_3$				

Based on the experience of previous studies, we selected the optimal synthesis conditions of 65 °C and a 1 h reaction time as the determined conditions. The article discusses the effects of four main influencing factors—namely, chitosan/acrylic acid concentration ratio, initiator concentration, acrylic acid neutralization degree and crosslinker concentration—on the water absorption properties of the synthesized highly absorbent gels through orthogonal tests. It can be seen from Table 3 that among the four factors affecting the water absorption rate, the degree of acrylic acid neutralization was the most important factor, the ratio of chitosan and acrylic acid monomer was the second most important factor, and the initiator had the least effect on the water absorption rate.

(2) Univariate analysis of the thermal preparation conditions of chitosan-acrylic acid highly absorbent hydrogels.

  (a) Effect of chitosan and acrylic monomer ratios on water absorption properties.

As can be seen from Figure 6a, when the monomer dosage was low, there were not enough monomers to graft and cross-link with chitosan. The induction period of the polymerization reaction was prolonged, the reaction rate was slow, and the cross-link density was low. The soluble part of the gel increased, so the absorption rate was low. As the monomer dosage increased, more available monomer molecules near the chitosan large radical chain growth site may have been grafted onto the chitosan backbone, which improved the grafting efficiency and the hydrophilicity of the corresponding highly absorbent agent which, in turn, improved the water absorption rate. When the monomer dosage reached a certain level, the monomer concentration was too high, and the increase in this ratio led to the self-polymerization of acrylic acid, and the grafting or linking of chitosan with acrylic acid. When the monomer concentration was too high, the increase of the ratio led to the fierce competition between the self-polymerization of acrylic acid and the grafting or linking of chitosan and acrylic acid, and thus the polymerization process had difficulty in dissipating heat. It was easy to cause explosive polymerization, but the polymerization reaction was not easy to control, and the chance of a self-cross-linking reaction in the specimen increased so that the cross-linking was too large. Thus, the water absorption multiple was reduced. Therefore, a more suitable acrylic acid volume to chitosan mass ratio was 6 mL/g, and the water absorption rate was 476 mL/g.

(b)　Effect of cross-linking agent dosage on water absorption performance.

The amount of cross-linking agent was one of the most important factors affecting the water absorption performance in the process of synthesizing the highly absorbent resins, because the amount of cross-linking agent directly affected the water absorption rate and water retention capacity of the system. As can be seen from Figure 6b, when the amount of cross-linking agent was small, the degree of cross-linking of the resin was small. With the addition of the cross-linking agent, the concentration of the cross-linking degree increased, and the water absorption capacity of the resin also increased. When the amount of cross-linking agent was too large, the higher concentration of cross-linking agent led to more cross-linking points being produced in the polymer network, and the free space between the polymer chains was reduced. The cross-linking points in the cross-linking network structure of the highly absorbent resin increased, and the cross-linking density increased, resulting in the polymer network space becoming smaller, the swelling becoming worse, and the water absorption rate decreasing. Thus, the amount of cross-linking agent should be moderate.

(c)　Effect of initiator dosage on water absorption performance.

The amount of initiator was an important factor affecting the gel performance. As can be seen from Figure 6c, when the amount of initiator was too small, the reaction system had fewer active centers, slower polymerization, lower cross-linking, and lower water absorption. With the increase of initiator dosage, the degree of cross-linking of the system gradually increased and the water absorption rate also increased; however, when the initiator dosage was too much, the active center of the system increased. This caused the cross-linking reaction: the cross-linking degree increased and the network structure of the system became tight, which is not conducive to water absorption, and thus the water absorption rate decreased. At the same time, when the initiator content was too much, the higher number of free radicals in the reaction system, the faster reaction speed, and more chain termination reactions also reduced the molecular weight of the polymer, and the contraction of the resulting cross-linked network decreased its liquid absorption rate.

(d)　Effect of neutralization degree of acrylic acid on water absorption performance.

The effect of neutralization degree on superabsorbent hydrogels was mainly distributed in three aspects. One was the ratio of -COOH and -COONa$^+$ in the system. Second, the neutralization degree also had a great influence on the polymerization rate. When the neutralization degree was low, the polymerization rate is fast. the degree of chain transfer was large, and the reaction was not easy to control. It was easy to burst polymerization, and the chance of self-cross-linking and low molecular weight acrylic resin in the final resin was large, which was not conducive to the improvement of the water absorption multiplier. When the neutralization degree was too high, the polymerization rate was slow and the conversion rate was low. Thirdly, when the neutralization degree exceeded a certain value, chitosan started to hydrolyze, and the cross-linked structure was destroyed. With the gradual increase of the neutralization degree, the degree of destruction increased and the water absorption rate decreased rapidly.

The effect of the degree of neutralization of acrylic acid was also investigated, while the other variables were kept at the following conditions: reaction temperature, 80 °C; reaction time, 1 h. The ratio of the volume of acrylic acid to the mass of chitosan was 3 mL/g. The effects of neutralization degree on the reaction yield and water absorption of the products are shown in Figure 6d. The results showed that the effects of the neutralization degree on the reaction yield and water absorption were similar. The reaction yield and water absorption increased with the increase of neutralization degree, reaching a maximum of 80% neutralization degree, which then decreased with the further increase of the neutralization degree. The reaction yield and water absorption increased with the increase of the neutralization degree, reaching a maximum of 80% neutralization degree, which then decreased with the further increase of neutralization degree. The increase in carboxylic acid anions enhanced the interchain repulsion, and thus the dynamics of the polymer network extension. As a result, the yield and water absorption increased. However, when excess NaOH was added to the reaction solution, the excess Na cations shielded the charge of the carboxylate anions; thus, preventing the effective repulsion between the carboxylate anions and decreasing the yield and water absorption. Therefore, a more suitable degree of acrylic acid neutralization should be 80%.

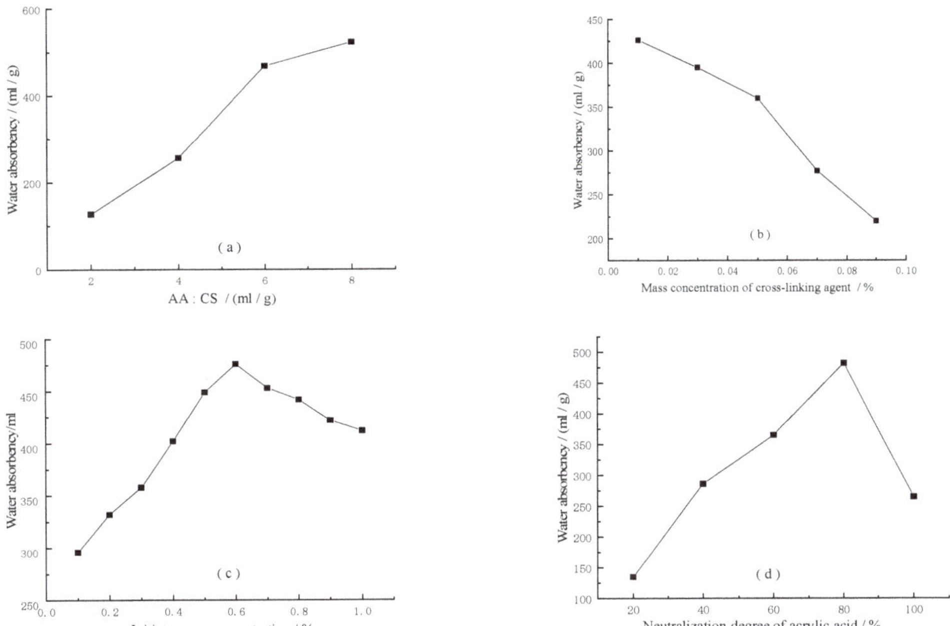

Figure 6. Trends in indicators versus factors, (**a**) chitosan and acrylic monomer ratios, (**b**) cross-linking agent dosage, (**c**) neutralization degree of acrylic acid, (**d**) neutralization degree of acrylic acid.

From the Figure 6, it can be seen that the water absorption multiplier tended to increase with the increase of chitosan mass fraction, and then gradually decrease. The water absorption multiplier tended to decrease with the increase of crosslinker concentration, and the water absorption multiplier tended to increase with the increase of chitosan/acrylic acid concentration ratio and initiator concentration. This is because in the graft copolymerization reaction, the chitosan graft acrylic monomer opportunity increased, to some extent, with the increase of chitosan mass fraction; however, when the chitosan mass fraction increased to a certain value, the water absorption multiplicity decreased with the decrease of the number of hydrophilic groups -COOH and -COO-. When the water absorption multiplier and the cross-link density of the polymer had a great relationship, cross-link density increased. The water absorption multiplier decreased due to the short distance between the cross-link points and the large elastic contraction of the cross-link network. When the amount of initiator additive was small, the decomposition rate was slow. The chain initiation reaction was slow. The grafting rate was low within a certain reaction time, and the water absorption multiplier of the synthesized highly absorbent gel was low.

3.3. Performance Test of Superabsorbent Hydrogel Foam Stabilizer

The above synthesized gel–foam stabilizer and acrylamide conventional foam stabilizer were added to the compound foaming agent (CAB:SDS was 6:4, concentration 0.6%) in a certain ratio, and stirred for 1 min using the Warning blender stirring method. Then, the foam was poured into a 2000 mL measuring cylinder, and the time required to reduce the foam volume by half was recorded for evaluating the foam stability. Experimental results are shown in Table 4 below.

Table 4. Test results of foam stabilization performance.

Number	Compound Foaming System	Stabilizer System/%		Foam Volume/mL	Half-Life of Defoaming/Min
1			0.1	980	60
2			0.3	950	80
3	SDS-0.24%	Acrylamide	0.5	890	110
4	CAB-0.36%		0.7	780	170
5			0.9	700	240
6			0.1	1280	230
7			0.3	1230	720
8	SDS-0.24%	Synthesis gel	0.5	1150	1080
9	CAB-0.36%		0.7	1060	780
10			0.9	980	595

As can be seen from the Table 4, The addition of foam stabilizer and the increasing concentration of foam stabilizer have a certain degree of influence on the foaming ability of surfactant. The main manifestation is that the foaming volume decreases to varying degrees compared with that before the addition of foam stabilizer. With the increasing concentration of foam stabilizer, the foam volume decreases. After the addition of two substances, the foaming volume decreased significantly. But it can be seen that the effect of synthetic gel on the foaming capacity of the system was smaller than that of acrylamide, which can be seen. This is because both of them achieve the purpose of stabilizing bubbles by thickening the solution. Increasing the viscosity of the mixed solution enhances the intermolecular force, which is convenient for the contradictory foaming mechanism of surfactant. Therefore, although these two foam stabilizers improve the foam stability to a certain extent, they inhibit the foaming ability of the system. So we need to find the balance between foam stabilization time and foam volume. The foam stabilization time of the synthesized gel is significantly longer than that of acrylamide.

As can be seen from the Figure 7, with the increase in the concentration of stabilizer, the gel–foam system foaming volume was reduced and the precipitation was reduced. The gel–foam system defoaming half-life showed a trend of first increasing, and then decreasing. At the same concentration, the synthesized highly absorbent gel showed a better foam stabilization effect compared with acrylamide, and its concentration of 0.5% showed the longest stabilization time of 1080 min, which was selected as a foam stabilizer for the foam compounding system.

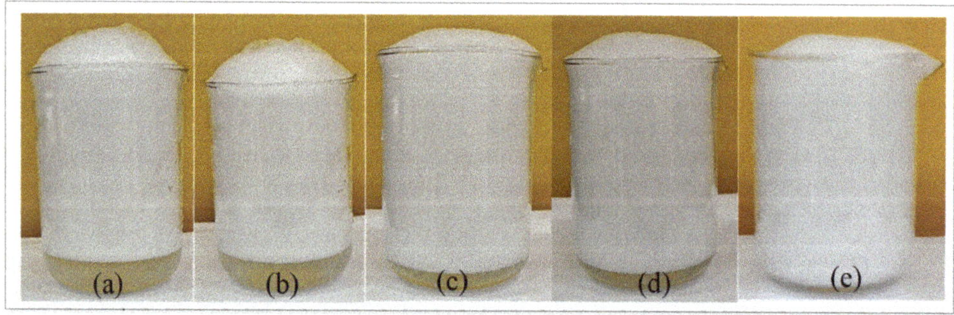

Figure 7. Gel–foams with different mass fractions of foam stabilizers ((**a**)—0.1%; (**b**)—0.3%; (**c**)—0.5%; (**d**)—0.7%; (**e**)—0.9%).

3.4. Heat Resistance Time of Gel–Foam

To study the temperature resistance of the gel–foam at different gel ratios, the experiment fixed the foaming agent (CAB:SDS of 6:4 and concentration of 0.6%) constant and added 2%, 4%, 6%, 8% and 1% of gel–foam, respectively. The half-lives were recorded by putting them into a drying oven at 200 °C.

Table 5 shows that the temperature resistance of gel–foam was higher than that of ordinary two-phase foam, and that it increased significantly with the increase of the mass concentration of stabilizer. This is because the main component of the ordinary foam film was water. As the temperature increased, the solution expanded and the thermal movement of molecules intensified, resulting in weaker intermolecular forces and less hydration. The surfactant molecules were not closely arranged, the solution viscosity was reduced, the drainage rate was accelerated, and the foam stability was reduced; therefore, the temperature resistance time was shorter. Due to the addition of the thickening agent in the gel–foam, the cross-linking reaction between molecules formed the three-dimensional network structure of the foam. The free water was effectively bound and fixed in the formed three-dimensional network structure, greatly improving the stability. As a result, the gel–foam had better temperature resistance.

Table 5. Temperature resistance time test.

Gel Ratio	0.1%	0.2%	0.4%	0.6%	0.8%	1%
Time (s)	180	420	660	840	960	1140

4. Application of Gel–Foam in Coal Gangue Hill Fire Area Treatment

In view of the fire situation of Yuquan coal gangue hill, gel–foam was used to treat a gangue hill.

4.1. Coal Gangue Mountain Measuring Point Arrangement

A coal gangue in Yu County, Yangquan, Shanxi, as shown in Figure 8, of 1 # coal gangue mountain preliminary determination, 1 # surface anomaly area, and 2 # surface anomaly area, as shown in Figure 9, was selected within the length of 35 m, width of 5 m and the length of 55 m, with a width of a 5 m long area for grid division. The arrangement of a high-temperature regional surface temperature and harmful gas detailed measurement points is seen below, with the same direction of each adjacent measurement point spacing for 5 m.

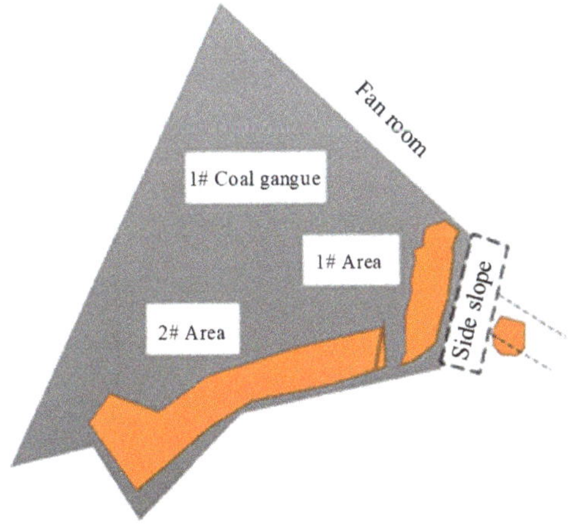

Figure 8. Coal gangue hills.

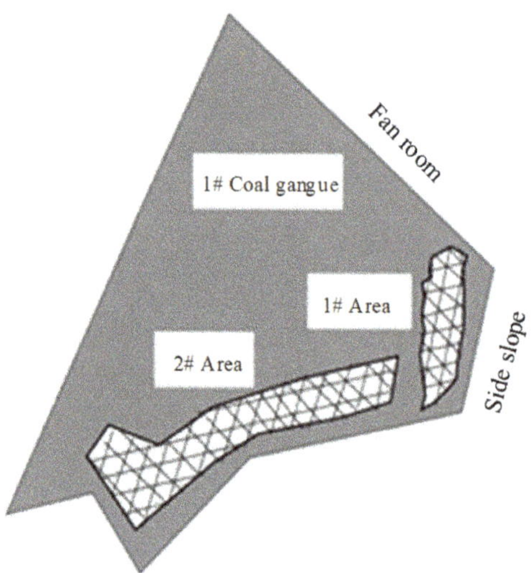

Figure 9. Grid division of surface anomaly area.

In the coal gangue mountain platform, a vehicle-mounted drilling rig is used for drilling, with a detection drilling depth according to the actual situation at the site. Generally, it is appropriate to see the gangue after continuing to drill 2 m, with a drilling diameter of 75 mm, and any two adjacent drill hole distances between 10 m: the specific layout is shown in Figure 10. There is an area using armored temperature sensors to measure and monitor the temperature of the bottom of the borehole, to record the temperature of each measurement point (°C), as seen in Figure 11, a coal gangue mountain surface temperature isogram.

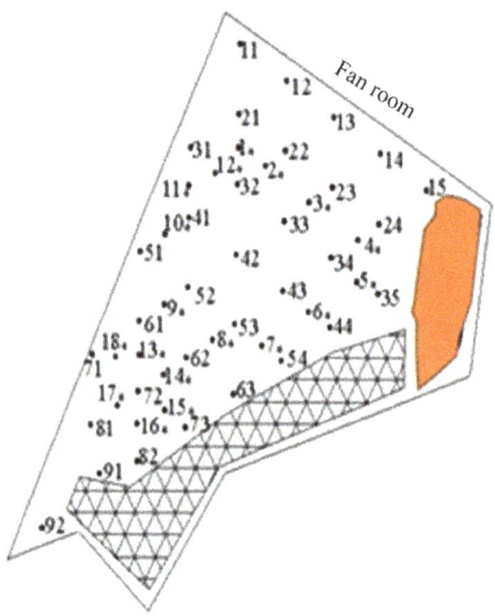

Figure 10. Distribution map of borehole detection points in coal gangue hill.

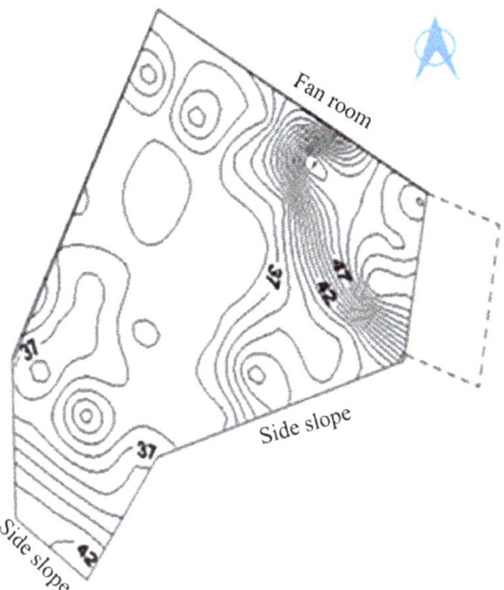

Figure 11. Surface temperature equivalence of coal gangue hill.

4.2. Gangue Hill Fire Prevention Technology

Using gel–foam fire-extinguishing technology through the construction of boreholes in the gangue mountain natural fire area, composite materials can be injected. These composite materials contain water, can absorb heat, can quickly reduce the temperature, and can effectively play the role of blocking the air leakage channel, achieving the purpose of extinguishing the fire area.

(1) Layout of glue injection borehole.

The arrangement of the grouting borehole is the same as the grouting borehole in Figure 10.

(2) Composite Colloid Materials and Formulas.

The composite colloidal material in the prevention and control of spontaneous combustion of gangue hill was prepared by a foaming agent, water, and a small amount of superabsorbent foam stabilizer.

4.3. Effect Test of Spontaneous Combustion Fire Control

(1) Surface temperature monitoring of gangue hill.

After the fire-extinguishing work, the surface temperature of the gangue hill was measured in a wide range by a portable infrared thermometer. The key monitoring area was the area near the slope, and the treated area. At the same time, the surface temperature around the gangue hill was recorded and compared. Since the surface temperature measurement position was not a scatter distribution, the data are no longer listed here. After 1, 3, 6, 10, 20, 30 days of monitoring, no surface high temperature area was found.

(2) Internal temperature monitoring of gangue hill.

In the process of pre-construction drilling and fire source excavation, a number of temperature measuring points in the internal fire zone were retained, and the positions of these measuring points were counted, as shown in Figure 10. There were 21 random measuring points in total. The internal temperature was measured, as shown in Figure 12.

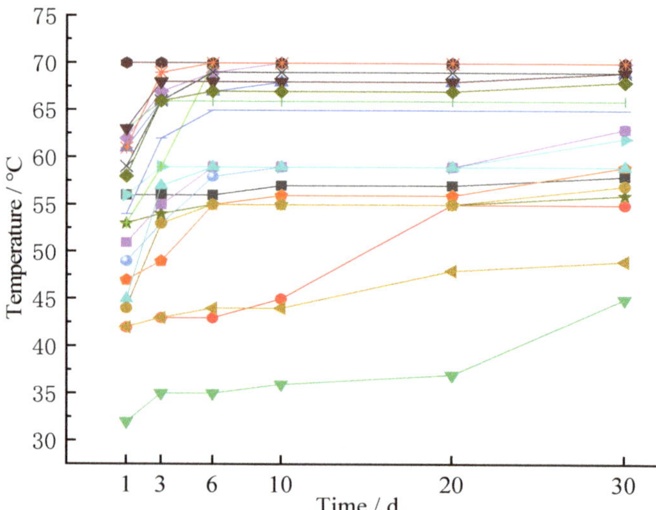

Figure 12. Monitoring results of internal temperature of gangue hill after fire-extinguishing.

From the above monitoring data, the surface temperature of the gangue hill was consistent with the surface temperature outside the gangue hill, and there was no high temperature area on the surface. The internal temperature of the gangue hill was up to 70 °C, which meets the requirements of coal-mine gangue hill disaster prevention and control norms. It can be said that the fire-extinguishing work of the gangue hill achieved good results. There was no re-ignition phenomenon in a month.

5. Conclusions

(1) With the increase of surfactant concentration, the foaming volume of the foam increased rapidly. When the surfactant concentration increased to a certain extent, the foam volume tended to be flat or slightly reduced;

(2) The foaming volume of the compound system increased, indicating that the compound surfactant had a synergistic effect, and the compound surfactant had better performance than the single surfactant;

(3) Among the four factors affecting the water absorption rate of foam stabilizers, the degree of acrylic acid neutralization was the most important factor, the ratio of chitosan and acrylic acid monomer was the second most important factor, and the initiator had the least effect on the water absorption rate;

(4) With the increase of foam stabilizer concentration, the foaming volume of the gel–foam system decreased, and the defoaming half-life of the gel–foam system first increased, and then decreased. At the same concentration, the synthesized superabsorbent hydrogel showed a good foam stabilization effect. It was selected as foam stabilizer for a foam compound system;

(5) A coal gangue mountain was drilled with gel–foam to extinguish a fire. Field use showed that the foamed gel could extinguish coal ignition effectively.

Author Contributions: Conceptualization, X.Z.; methodology, X.Z.; software, X.Z.; validation, X.Z.; formal analysis, Y.P.; investigation, Y.P.; data curation, Y.P.; writing—original draft preparation, Y.P.; writing—review and editing, X.Z.; project administration, X.Z.; funding acquisition, X.Z. All authors have read and agreed to the published version of the manuscript.

Funding: This work was sponsored by National Natural Science Foundation of China (No51804270), the Natural Science Foundation of Hunan Province (No2020JJ5547) and the Research Fund of Hunan Provincial Department of Education (No19C1745).

Institutional Review Board Statement: Not applicable.

Informed Consent Statement: Not applicable.

Data Availability Statement: Not applicable.

Conflicts of Interest: The authors declare no conflict of interest.

References

1. Zhou, F.-B.; Shi, B.-B.; Cheng, J.-W.; Ma, L.-J. A new approach to control a serious mine fire with using liquid nitrogen as extinguishing media. *Fire Technol.* **2015**, *51*, 325–334.
2. Cheng, W.; Hu, X.; Xie, J.; Zhao, Y. An intelligent gel designed to control the spontaneous combustion of coal: Fire prevention and extinguishing properties. *Fuel* **2017**, *210*, 826–835. [CrossRef]
3. Guo, Q.; Ren, W.; Zhu, J.; Shi, J. Study on the composition and structure of foamed gel for fire prevention and extinguishing in coal mines. *Process Saf. Environ. Prot.* **2019**, *128*, 176–183. [CrossRef]
4. Wang, G.; Yan, G.; Zhang, X.; Du, W.; Huang, Q.; Sun, L.; Zhang, X. Research and development of foamed gel for controlling the spontaneous combustion of coal in coal mine. *J. Loss Prev. Process Ind.* **2016**, *44*, 474–486. [CrossRef]
5. Xue, D.; Hu, X.; Cheng, W.; Wei, J.; Zhao, Y.; Shen, L. Fire prevention and control using gel-stabilization foam to inhibit spontaneous combustion of coal: Characteristics and engineering applications. *Fuel* **2020**, *264*, 116903. [CrossRef]
6. Wu, M.; Liang, Y.; Zhao, Y.; Wang, W.; Hu, X.; Tian, F.; He, Z.; Li, Y.; Liu, T. Preparation of new gel foam and evaluation of its fire-extinguishing performance. *Colloids Surf. A Physicochem. Eng. Asp.* **2021**, *629*, 127443. [CrossRef]
7. Guo, Q.; Ren, W.; Bai, L. Properties of foamed gel for coal ignition suppression in underground coal mine. *Combust. Sci. Technol.* **2019**, *191*, 1294–1308. [CrossRef]
8. Ren, W.; Guo, Q.; Zuo, B.; Wang, Z. Design and application of device to add powdered gelling agent to pipeline system for fire prevention in coal mines. *J. Loss Prev. Process Ind.* **2016**, *41*, 147–153. [CrossRef]
9. Xi, X.; Shi, Q. Study of the preparation and extinguishment characteristic of the novel high-water-retaining foam for controlling spontaneous combustion of coal. *Fuel* **2021**, *288*, 119354. [CrossRef]
10. Shi, Q.; Qin, B. Experimental research on gel-stabilized foam designed to prevent and control spontaneous combustion of coal. *Fuel* **2019**, *254*, 115558. [CrossRef]
11. Li, S.; Zhou, G.; Wang, Y.; Jing, B.; Qu, Y. Synthesis and characteristics of fire-extinguishing gel with high water absorption for coal mines. *Process Saf. Environ. Prot.* **2019**, *125*, 207–218. [CrossRef]
12. Xiao, C.; Balasubramanian, S.N.; Clapp, L.W. Rheology of Viscous CO_2 Foams Stabilized by Nanoparticles under High Pressure. *Ind. Eng. Chem. Res.* **2017**, *56*, 8340–8348. [CrossRef]
13. Wang, X.; Mohanty, K. Improved Oil Recovery in Fractured Reservoirs by Strong Foams Stabilized by Nanoparticles. *Energy Fuels* **2021**, *35*, 3857–3866. [CrossRef]
14. Verma, A.; Chauhan, G.; Ojha, K.; Padmanabhan, E. Characterization of Nano-Fe_2O_3-Stabilized Polymer-Free Foam Fracturing Fluids for Unconventional Gas Reservoirs. *Energy Fuels* **2019**, *33*, 10570–10582. [CrossRef]
15. Wasan, D.T.; Nikolov, A.D.; Aimetti, F. Texture and stability of emulsions and suspensions: Role of oscillatory structural forces. *Adv. Colloid Interface Sci.* **2004**, *108–109*, 187–195. [CrossRef] [PubMed]
16. Ravera, F.; Santini, E.; Loglio, G.; Ferrari, M.; Liggieri, L. Effect of Nanoparticles on the Interfacial Properties of Liquid/Liquid and Liquid/Air Surface Layers. *J. Phys. Chem. B* **2006**, *110*, 19543–19551. [CrossRef] [PubMed]
17. Zhang, S.; Lan, Q.; Liu, Q.; Xu, J.; Sun, D. Aqueous foams stabilized by Laponite and CTAB. *Colloids Surf. A Physicochem. Eng. Asp.* **2008**, *317*, 406–413. [CrossRef]
18. Shao, Z.L.; Wang, D.M.; Wang, Y.M.; Zhong, X.; Tang, X.; Hu, X. Controlling coal fires using the three-phase foam and water mist techniques in the Anjialing Open Pit Mine, China. *Nat. Hazards* **2015**, *75*, 1833–1852. [CrossRef]
19. Chen, Z.; Wei, X.L.; Li, T. Experimental Investigation on Flame Characterization and Temperature Profile of Single/Multiple Pool Fire in Cross Wind. *J. Therm. Sci.* **2021**, *30*, 324–332. [CrossRef]
20. You, Q.; Wang, H.; Zhang, Y.; Liu, Y.; Fang, J.; Dai, C. Experimental study on spontaneous imbibition of recycled fracturing flow-back fluid to enhance oil recovery in low permeability sandstone reservoirs. *J. Petrol. Sci. Eng.* **2018**, *166*, 375–380. [CrossRef]
21. Wassmuth, F.R.; Hodgins, L.H.; Schramm, L.L.; Kutay, S.M. Screening and coreflood testing of gel foams to control excessive gas production in oil wells. In Proceedings of the SPE/DOE Improved Oil Recovery Symposium, Tulsa, OK, USA, 3–5 April 2000.

MDPI

St. Alban-Anlage 66

4052 Basel

Switzerland

Tel. +41 61 683 77 34

Fax +41 61 302 89 18

www.mdpi.com

Energies Editorial Office

E-mail: energies@mdpi.com

www.mdpi.com/journal/energies